Transforming Data into Informed Decisions across Clinical and Non-Clinical Domains

István Fekete

Transforming Data into Informed Decisions across Clinical and Non-Clinical Domains

Springer

István Fekete
Amity Institute of Psychology and Allied Sciences
Amity University Kolkata
Kolkata, West Bengal, India

ISBN 978-3-032-02594-4 ISBN 978-3-032-02595-1 (eBook)
https://doi.org/10.1007/978-3-032-02595-1

This Springer imprint is published by the registered company Springer Nature Switzerland AG
The registered company address is: Gewerbestrasse 11, 6330 Cham, Switzerland

To my mentors and parents, whose support enabled me to write this book. This book is dedicated to researchers and my students, whose enthusiasm and curiosity have motivated me to learn more and expand our knowledge of a sophisticated, enhanced, and more reliable data analysis.

Preface

The purpose of this book is to investigate one of the relatively recent advances in the field of statistics. This book published by Springer is of value to anyone who wishes to expand their understanding of the so-called Minimal Clinically Important Difference (MCID). In addition to offering the theoretical background of MCID, this book discusses the real-world difficulties faced by researchers and highlights some inconsistencies in calculating them. The purpose of this book is to encourage, inform, and stimulate the use of MCID across a range of scientific fields including the humanities, with the latter being a novelty aspect of the book. This book is dedicated especially to students of humanities who pursue quantitative research and to their professors who I wish they could integrate this metric in their curriculum.

The concept of MCID has long been established in clinical research as compared to nonclinical fields since the 1980s. MCID is a threshold that quantifies mainly the clinical and patient-centric importance of treatment effects. In other words, the emphasis is on patient-relevant change or difference. This book explores the potential of MCID beyond healthcare, examining its relevance in diverse fields such as psychology, biology, education research, business, economics, finance, humanities, linguistics, the social sciences, and drug design inter alia.

This book was inspired by the need for meaningful thresholds or cut-offs that enhance decision-making based on statistically significant results. MCIDs improve the quality and reliability of scientific publications but also research and technical reports (also in the industry) across various fields outside the clinical and the scientific domain.

Hence, the major contribution of the book is the adaptability of MCID in nonclinical fields such as the social sciences and the humanities. The book begins with an overview of the theoretical background of MCID, followed by chapters that discuss the applicability of MCID in nonclinical domains. These sections present case studies, examples, and empirical research demonstrating as to how MCID can be utilized in those fields.

I am grateful to my mentors, and to the team at Springer Nature, for their support throughout writing this book. It is my hope that this book will serve as a resource for research scholars, practitioners, and data scientists among other professionals by giving them new ideas about the enhancement of their research findings.

Kolkata, West Bengal, India Istvan Fekete

Ethics Approval The author confirms that no ethical approval was required for the research or presentation of data in this book, because writing the book did not involve any experimentation on human participants or animals, as the book is based on literature review, theoretical work, and secondary data analysis.

Competing Interests

The author declares that he has no competing financial or nonfinancial interests to disclose, or conflicts of interest to report. The author does not have any affiliations with any organization or entity with any financial interest.

About the Author

I completed my PhD in psycholinguistics in 2013 in Budapest, Hungary, which laid the foundation for a decade-long post-PhD work as a data-analyst with diverse data types, from psycholinguistics datasets through educational datasets to those encompassing mainly medical, psychological, and biological domains. During my post-PhD years, I observed frequently how large datasets can magnify/amplify trivial (i.e., small) into *statistically (even) highly significant* ones, despite the actual magnitude of these differences often being small, and thus, rendering them practically and scientifically irrelevant; a conundrum that has troubled me when analyzing data, and most importantly, making important decisions unreliable.

I have observed that, at the same time, other even more experienced colleagues and PhD scholars were unaware of this issue; they just cared about statistical significance and were doing so-called "p-hacking" as they say nowadays: they increased the sample size until the desired effect became statistically significant, which is a statistical-methodological malpractice. Such an unethical practice can lead to false conclusions even if the result is significant. Another problem is that in most cases sample-size calculations are not performed. Sample-size calculation, namely, could safeguard against "p-hacking." Just check for yourself how many publications have been published in which sample-size calculation was performed, or, at least, a justification of sample-size.

These observations motivated me to look for a metric capable of filtering out these statistically significant though trivial/negligible differences or changes from baseline. This inspired me to start writing this book. Second, it appeared to me that such a metric has remained unrecognized and completely underemployed among various fields beyond the clinical field including the business industry, the social and political sciences, the humanities (including my original field of linguistics and psycholinguistics, specifically), or biology to name a few domains.

This book also encapsulates my career in the realms of Health Technology Assessment (HTA) when I worked as a senior consultant and biostatistician analyzing clinical trials for early drug benefit assessment at a company named SmartStep Consulting GmbH in Hamburg, Germany. Here, I had the opportunity to apply

Minimal Clinically Important Difference (MCID). I was sure at that time that there existed no such metric, for example, in linguistics studies.

Professionals in every scientific field have to make important decisions based on statistics. Without this metric their decisions might be unreliable or even misleading. I firmly believe that such a metric could enhance informed decisions and that it should be introduced in every scientific field. If wrong or suboptimal decisions are made based on trivial differences, it might have drastic consequences: professionals might over- or underreact certain changes irrespective of scientific field. Just think of the stock market or the medical field.

I concluded that most researchers just employed effect size measures at best, beyond traditional statistical significance testing. My role as a senior biostatistician allowed me to leverage MCID in my own work later too, thereby refining the precision of statistical findings by employing MID (Minimal Important Difference) to discern the subtle yet meaningful effects of statistically significant differences or changes.

Statistical significance is different from *clinical significance*, with the latter concept being patient-relevant and the former being "just" data-driven in a lab-setting. For example, in pain management, the distinction between a *statistically significant improvement* and a *clinically noticeable impact* on a patient's pain can be profound and "miles away from each other." While, for instance, a 20% reduction in pain may be deemed even *statistically highly significant* in research settings (e.g., in publications or what I earlier called lab-setting), it often falls short of the substantial relief needed for meaningful quality of life (QoL) improvements for patients (e.g., in hospital settings or in HTA). Note that demographic data are excluded from this argumentation intentionally, because some MID-metrics are not personalized at the moment. For a personalized metric (MID), one would need a model which incorporates MID. At the moment, in most cases, patient characteristics are excluded from the calculation of MID. At times, researchers employ stratification and compute the MID for each stratum separately at best, which I deem as a temporary and optimal solution.

To exemplify, a study may report that a new pain management therapy achieves a statistically significant 20% reduction in pain scores on an 11-point pain scale compared to standard-of-care, abbreviated as SoC, of pain management. However, for patients living with, for example, chronic pain, such as those with severe osteoarthritis (OA) or neuropathic pain conditions, a mere 20% reduction might not mean a noticeable, meaningful, or substantial improvement in their daily functions or overall well-being and QoL. Again, we can observe that patient demographics are excluded, which will be addressed multiple times in the book.

Finally, in my role as a full professor in Research Methods and Statistics, I have observed that the concept of MCID/MID is absent from the curriculum in many disciplines in the World (such as psychometrics, statistics for psychologists, advanced statistics, econometrics, mass communication, etc.). Those seven universities where I have taught so far, the metric is not part of the curriculum, not even in advanced statistics courses and biostatistics for PhD scholars. This omission is regrettable, given that MCID/MID as a patient-relevant metric that influences sometimes life-saving decisions.

I am going to include examples from already published SCOPUS-articles from various fields that showcase *statistically highly significant but practically irrelevant differences or changes*. Also, I am going to incorporate examples from *contradictory HTA decisions* (i.e., contradictory efficacy of medical products) from different countries in which early drug benefit assessment was based on identical datasets from clinical trials.

This book is intended mainly for professionals working or interested in nonclinical field, such as social sciences, where they do not employ MCID/MID. They could include this metric in their statistical repertoire and also as sensitivity analysis.

My secondary aim is to introduce this metric in the curricula (quantitative research, research methods, statistics) in the social sciences and the humanities to enhance the reliability and quality of research findings. Data is the new gold, as they say; in an era of big data, the humanities also need quantitative research and statistics. Just think of corpus-linguistics, psycholinguistics, questionnaire studies in socio-linguistics, etc.

Acknowledgments

First and foremost, I would like to chronologically acknowledge Professor Dr. Csaba Pléh, whose inspirational involvement in 2006 in Budapest, Hungary, in sparking my love for experimental linguistics has been crucial to my scientific career. I began my academic career under his scholarly direction. Professor Dr. Gerd Hentschel in Oldenburg, Germany, helped me further with my academic endeavors during my post-doctoral phase. His guidance impacted my personal growth in addition to helping me in my academic endeavors. Third, my career outside of academia was greatly influenced by Dr. Timm Volmer (Hamburg, Germany). He and his team introduced me to the Health Technology Assessment (HTA), for which I am very grateful. For our conversations on econometrics, I am grateful to Prof. Dr. Sakhi Roy. The editorial assistance of Springer Nature and the direction of Dr. Alina Cepraga were invaluable.

I extend my gratitude to Amity University Kolkata, for generously providing patience, time, and scholarly freedom that supported my research and writing endeavors, especially to Prof. (Dr.) Akash Kumar Mahato, and to Honorable Vice Chancellor of Amity University Kolkata, Prof. (Dr.) Ashok Srivastava, and highly respected Senior Vice President, U. Ramachandran. Last but not least, I am indebted to the support of my parents. The collective influence of these individuals has been a great encouragement. The errors in the book are mine.

About the Book

The unique aspect of this book is that it integrates Minimal Clinically Important Difference (MCID) and Minimal Important Difference (MID) among similar metrics across a wide array of disciplines outside the traditional clinical and psychological fields, hitherto unaccomplished comprehensively in its present form in the social sciences and the humanities (such as the educational sciences, environmental studies, even business analytics and finance, linguistics, economics, or biology inter alia).

For example, in speech therapy, an MCID-equivalent measure could assess the smallest change in *fluency* of a child (as the outcome variable) that makes a meaningful and perceivable difference to individuals with speech impairments, to their mother, relatives, nurses, teachers, care-givers, or the individuals in their immediate environment (sometimes referred to as the signaling environment or signal system) and not just a statistically significant difference or change that the majority of studies report.

To simplify the problem, what if in an experiment there is a statistically significant increase in *vocabulary size* or *fluency* of this child, measured by some standardized test quantitatively or qualitatively, but the people in the immediate environment of the child does not recognize or perceive it? Unless we know the reason or reasons (shyness, for example), we have to assume that the child has not crossed the MCID. He/she needs more practice, vocabulary extension, etc. This is where "statistical significance" and "practical significance" diverge, rendering the topic of this book. We can call these practical improvements in the two linguistics examples as *linguistic relevance*.

Let me continue with another example from the field of biology after two examples from linguistics in which an MCID-like metric (we can call it MID; Minimal Important Difference) could *determine the minimum population size increase or habitat improvement that would be considered biologically relevant for the survival and recovery of endangered species and, again*, not just a statistically significant difference or change.

In ecological restoration projects, again, an MCID-like measure could assess the *minimum change in pollution levels*, *habitat quality*, *air quality level* (AQI), or species diversity that is necessary to have a meaningful positive and clearly perceivable impact on the ecosystem. By "clearly perceivable impact," I mean an effect that is relevant in the subject area, to the people, or to the healthcare system (the latter in terms of both costs and clinical variables).

This multidisciplinary approach of the book addresses a significant methodological research gap by demonstrating the utility of MCID/MID in facilitating enhanced and more reliable decision-making processes in various clinical and nonclinical fields. By decision-making processes, I mean making an enhanced and informed decision based on statistically significant results, which have relevant, reliable, and practical implications (practical in the sense of noticeable and perceivable). You can think of this metric as supporting evidence for the practical implication of a research finding or as a decision whether to accept a significant result as relevant or not. This metric eases decision-making. A decision can be, for example, dosage escalation in a therapy, investment in a particular stock at the stock-market, or stopping a project at a certain point.

Furthermore, the book discusses methodological diversities across different Health Technology Assessment (HTA) systems in a few countries, underscoring controversial (!) real HTA decisions, a serious issue, often stemming from divergent national regulatory and policy frameworks across countries, showcasing another unique aspect of the book.

In addition, the book also highlights the tension between universal/standardized/singular MCID/MID contra context- and indication-specific cut-offs (i.e., disease-specific thresholds), which is a still ongoing debate within the medical community. This tension is especially there in the German HTA system called AMNOG (translated as "German Medicines Market Reorganization Act"). AMNOG came into force in 2011, with the purpose of evaluating new drugs and determining their price (the nomenclature of this process is "early benefit assessment").

Stakeholder perspectives, ranging from patients to policymakers, represent another unique feature. The lack of consensus across HTA bodies of various countries regarding methodological approaches for incorporating MCID/MID into HTA reflects broader issues within the field, a topic that adds extra complexity.

The book also proposes a forward-looking perspective on how MCID/MID can be integrated into remote healthcare among other newly emerging fields such as wearables (e.g., a smart watch or biosensors). The book addresses underexplored areas in MCID/MID, specifically the potential and hurdles (and not necessarily the need!) for global/universal HTA harmonization/standardization efforts, the integration of patient preferences, and the exploration of the application of MCID/MID in nonclinical domains, the unique and interdisciplinary aspect of the book.

This book is not just a compilation of knowledge but a call to action for MCID/MID-based enhanced, insightful, and impactful decision-making based on significant statistical results across various nonclinical disciplines. In nonclinical fields, such a metric does not exist yet, which leads to the acceptance of statistically highly significant but irrelevant/practically not meaningful results, which can lead to suboptimal decisions and Type I error.

The book can also be conceived of as a coursebook for the following courses (but not limited to): Advanced Statistics, Decision-Making in Medicine and Psychology, HTA, Research Methods and Statistics in Psychology, Biostatistics, Clinical Research Methods, Evidence-Based Medicine, Health Outcomes Research, Epidemiology, Patient-Centered Outcomes Research (PCOR), Quality of Life Research, Rehabilitation Sciences, Pharmacoeconomics, Rehabilitation Sciences, Gerontology and Aging Sciences, Behavioral Economics in Health, Public Health Policy and Public Health, Medical Ethics, Health Economics, Mathematical Statistics, Social Science Methodology, Sports Medicine and Exercise Science, Nursing Research and Practice, Pharmacy Practice and Research, Health Psychology, Social Work in Healthcare, Health Communication, Econometrics, Global Health and International Health Systems, Precision Medicine and Genomics, Causal Inference in Social and Health Sciences, Quality of Life Research, Social Science Methodology, and Personalized Medicine inter alia.

Contents

Abbreviations

ADME	Absorption, Distribution, Metabolism, and Excretion
AI	Artificial Intelligence
AMNOG	Act on the Reform of the Market for Medicinal Products
AOTMiT	Agency for Health Technology Assessment and Tariff System (Polish HTA)
AQI	Air Quality Index
ASA	American Statistical Association
ATMP	Access to Advanced Therapy Medicinal Products
BP	Blood Pressure
BZD	Benzodiazepine
CEA	Cost-Effectiveness Analysis
CMIE	Centre for Monitoring Indian Economy
COPD	Chronic Obstructive Pulmonary Disease
DCGI	Drug Controller General of India
DiD	Difference-in-Differences or DD
DILI	Drug-Induced Liver Injury
EBM	Evidence-Based Medicine
EC	European Commission
EC_{50}	Effective Concentration 50
EEG	Electrophysiology
FIH	First-in-Human Trials
G-BA	Gesameter Bundersausschuss (The Federal Joint Committee)
GCP	Good Clinical Practice
HAS	Haute Autorité de Santé (HTA body in France)
HED	Human Equivalent Dose
HEOR	Health Economics and Outcomes Research
HITAP	Health Intervention and Technology Assessment Program in Thailand
HRQoL	Health-Related Quality of Life
HTA	Health Technology Assessment
HTAi	Health Technology Assessment International
HED	Human Equivalent Dose

IC_{50}	Half Maximal Inhibitory Concentration
ICER	Institute for Clinical and Economic Review
IND	Investigational New Drug
IQWiG	The Institute for Quality and Efficiency in Health Care (in German: InstitutfürQualitätundWirtschaftlichkeitimGesundheitswesen,IQWiG)
JCA	Joint Clinical Assessment
KPI	Key Performance Indicator
LOAEL	Lowest Observed Adverse Effect Level
LD_{50}	Median Lethal Dose
LMICs	Low-and-Middle-Income countries
MAD	Multiple Ascending Dose
MCID	Minimal Clinically Important Difference
MCII	Minimal Clinically Important Improvement
MDC	Minimal Detectable Change
MID	Minimal Important Difference
MRSD	Maximum Recommended Starting Dose
MS	Multiple Sclerosis
MTD	Maximum Tolerated Dose
ML	Machine-Learning
NDA	New Drug Application
NHS	National Health Service
NICE	The National Institute for Health and Care Excellence (HTA body in England)
NOAEL	No Observed Adverse Effect Level
OA	Osteoarthritis
PCOR	Patient-Centered Outcomes Research
PFT	Pulmonary Function Test
PRO	Patient-Reported Outcome or Patient-Relevant Outcome
PROM	Patient-Related Outcome Measure
PSM	Propensity Score Matching
QALY	Quality-Adjusted Life Year
QoL	Quality of Life
RCT	Randomized Controlled Trial
REACH	Registration Evaluation Authorization and Restriction of Chemicals
ROI	Return of Investment
RT	Reaction Time
RWE	Real-World Evidence
SAD	Single Ascending Dose
SAP	Statistical Analysis Plan
SCB	Substantial Clinical Benefit
SD	Standard Deviation
SDM	Shared Decision-Making
SDG	Sustainable Development Goals
SE	Standard Error

SMA	Spinal Muscular Atrophy
SoC	Standard of Care (i.e., best practice therapy in a given country for a given indication)
TOEFL	Test of English as a Foreign Language
ULN	Upper Limit of Normal
UX	User Experience
WOMAC	Western Ontario McMaster Universities Osteoarthritis Index

Chapter 1
Introduction to Minimal Clinically Important Difference (MCID) and Minimal Important Difference (MID)

Abstract The chapter explains the concepts of minimal clinically important difference (MCID) and minimal important difference (MID), focusing on their relevance in differentiating between meaningful outcomes and mere statistical results. The book opens with a case study of India's air pollution. These examples illustrate the overreliance on p-values and statistical thresholds—ignoring whether the change is meaningful to the patient or public health. The hypothetical study used to illustrate the concept assumes a statistically significant reduction in hospitalization due to asthma because of an odd–even vehicle usage scheme. The chapter comments on how clinical relevance is forgotten in the drive to prove something "significant." Even with a minimal reduction of 0.5%, with a sufficiently large sample size, asthma hospitalizations could statistically "drop" during the odd–even scheme. However, this reduction would be imperceptible to the patients and irrelevant clinically and economically.

The chapter then provides a conceptual and historical synopsis of MCID and MID. Both MCID and MID emerged in the clinical literature in the 1980s. MCID sought to define the minimum clinically relevant change in treatment effect that a patient perceives as beneficial or harmful enough to warrant a change in management. Unlike the probability threshold of $p < 0.05$, MCID employs relevance, subjectivity, and tangible outcomes. Moreover, it sets itself apart from economic significance, which concerns itself with whether the effect size justifies the costs incurred in the context of health technology assessment (HTA).

To conclude, this chapter lays the groundwork for a book that intends to reposition MCID/MID from its clinical roots to include the more theoretical realms of psychology, economics, social science, and even the humanities. It advocates a change in one's thinking from the shallow obsession with p-values to a thoughtful shift toward meaningful change that is anchored in PROs, clinical insights, and repercussions beyond the clinical setting. The subsequent chapters will discuss multidisciplinary MCID/MID computation and interpretation methods, practical issues, and conceptual frameworks.

Keywords Minimal clinically important difference (MCID) · Minimal important difference (MID) · Clinical research · Patient care

I. Fekete, *Transforming Data into Informed Decisions across Clinical and Non-Clinical Domains*, https://doi.org/10.1007/978-3-032-02595-1_1

Introduction

I am writing this book in India. One of the biggest problems in India is air pollution and its impact (i.e., morbidity and mortality) on respiratory and other health outcomes in Delhi (e.g., Nagpure et al., 2014). India, with its vast population of more than 1.4 billion people in 2024 (e.g., Countrymeters, 2024; Worldometer, 2024) and rapid industrialization, faces significant challenges with air pollution, to the extent of estimated 7350–16,200 premature deaths and six million asthma attacks per year in the greater Delhi area (Guttikunda & Goel, 2013). Studies often investigate the short-term impact of air pollution interventions, such as the implementation of odd–even vehicle schemes, on respiratory health outcomes (e.g., Kumar et al., 2017).

Suppose a hypothetical study is conducted in Kolkata to evaluate the impact of a month-long odd–even vehicle scheme (where cars with odd and even number plates are allowed on the roads of Kolkata on alternate days) on the incidence of hospital visits for asthma-related medical conditions. Let us assume that this study finds a statistically significant (or even statistically *highly significant*!) reduction in air quality index (AQI)[1] during or after the intervention period compared to the previous month.

We would consider this research finding a scholarly (and, of course, political) triumph. Most probably, the answer is yes; it is "good" for the research team, for the people of Kolkata, the patients, the hospitals, and, last but not least, the politicians and stakeholders who initiated and financed the study. Yet, what we often see as significant results in statistics is where this book starts. The objective of this book is to start from statistically significant results and examine the *practical relevance* of such results.

The statistically significant reduction in hospital visits (number of hospitalizations, reduction in asthma attacks, duration of hospitalization in days, etc.), usually referred to as the "clinical endpoints" in the clinical field and in HTA, or the dependent variable (sometimes referred to as outcome measure, etc.) in psychology (Field, 2009), *which albeit statistically significant, might have a very small extent, magnitude, and effect size*—say, a reduction of 0.5% in hospital visits for asthma. So small that it is not perceivable at the clinical endpoint (e.g., improvement on the Pulmonary Function Test (PFT)). What is more, it might not even be perceivable to the patient subjectively. And, objectively, the PFT values would remain the same.

Note that if we increase the sample size sufficiently (which is considered a scientific malpractice), even a trivial difference will become significant with almost 100% certainty, a phenomenon that has been documented extensively (e.g., Brezina & Meyerhoff, 2014; Lin et al., 2013; Sullivan & Feinn, 2012). We are going to touch upon this topic multiple times when we talk about pre-specification of sample size, sample-size calculation, and Type I error (false positive results).

In this scenario of AQI, even though the result is *statistically significant*, the so-called *clinical relevance* or *clinical significance* of the observed reduction of, say,

[1] https://www.aqi.in/

0.5% is scientifically questionable for being too small or trivial. You should understand now why I underscored sample size previously. In other words, there might be a *statistically significant difference* (or *reduction*) that is *clinically not relevant*. But why? Because the threshold for considering an intervention effective in terms of public health might require a *much larger reduction* in hospital visits to offset the medical inconvenience or the *economic costs* (what is sometimes referred to as "economic significance"; see, e.g., Engsted, 2009).

Engsted (2009) differentiates between "statistical significance" and "economic significance" in the following way: "Economic significance" refers to the real-world implications or effects of statistical results, while statistical significance refers to statistical findings only. In HTA, economic significance and statistical significance are interconnected in some countries.

Returning back to our threshold of "clinical significance," this threshold in reduction, in AQI, for example, is called the minimal clinically important difference (MCID), first introduced by Guyatt et al. (1987) and Jaeschke et al. (1989), and sometimes also called minimal important difference (MID).

This book is about what we exactly mean by "a much larger reduction" and MCID, and as to how we can compute this threshold or cut-off. The primary and novel objective of this book is to introduce a metric analogous to the MCID/MID in nonclinical domains such as the social sciences, the business world, political surveys or the humanities, to name a few.

But how high or low should this MCID be, and how can we determine it in this illustrative example? How large should a "much larger reduction" be? Who decides on the threshold? What is "clinically not relevant"? What is patient relevance? What are patient-relevant outcomes (PROs)? I am going to use PRO as both referring to patient-relevant outcome and patient-related outcome, as they are very similar notions. How many MCIDs are there? Can MCID be individualized? Why are there so many of them? How do we calculate MID/MCID?

King (2011) differentiates between MCID, MID, MCII (minimal clinically important improvement), and MDC (minimal detectable change). Is there a standardized metric for MCID? What speaks for and against the standardization of MCID? I would like to leave these questions unanswered for the time being, because we are going to explore these and more in the upcoming chapters.

This example with the hypothetical air-pollution study underscores the importance of distinguishing between *statistical significance* (e.g., Cohen, 1988a, b) and *clinical significance* or *practical significance* (e.g., Jacobson et al., 1999; Kazdin, 1999). Note that there are multiple terms in the literature for the latter (e.g., King, 2011). Statistical significance just shows that an *observed difference between groups* or *observed change from baseline within a group* is not likely to have occurred by chance, according to a predefined probability threshold (which is usually 0.05, i.e., 5%), according to the classic source by Fisher (1925).

This book is on MID/MCID. Notwithstanding, King (2011) criticizes the concept of MCID. There are multiple names: for instance, the minimal important difference (MID), minimal clinically important difference (MCID), clinically important difference (CID), minimally detectable difference (MDD), and subjectively significant difference, all share the same domain of concern, thus presenting quite different meanings.

She (King, 2011) contends that "minimally detectable difference" and "minimum detectable change" also share these perplexing nuances. Often, definitions, no matter how intricate, do not impact the methods used to estimate these values. Patient rating of change (global transition items), clinical anchors, standard error of measurement (SEM), and effect size (ES) are the four methods primarily used to estimate MIDs (King, 2011).

King (2011) presents and critiques these methods. Even though it is preferable to consider an MID valid across uses, no such "universal MID" exists, consistent with my own opinion. In fact, it varies based on population and context. For any given PRO (scale), the MID is not a fixed attribute but is context-sensitive. Even at an aggregate level, the MID is sensitive to the clinical context, the specific decision at hand, the patient's baseline value, and whether they are on an improvement trajectory or not. Thus, specific estimates of MID should retain a degree of caution.

Let us turn away from MID/MCID for a moment. Wasserstein and Lazar (2016) provide a contemporary perspective on the (mis-)use and (mis-)interpretation of statistical significance and p-values. They discuss the guidelines of American Statistical Association (ASA) and the broader implications for statistical practice (Wasserstein et al., 2019). Similarly, Kline (2013) also discusses the foundational concepts around the debate on significance testing, highlighting the historical disputes and misunderstandings related to p-values. Some researchers even go as far as claiming to *abandon the p-value* (e.g., in corpus linguistics, e.g., Koplenig, 2019).

Definition and Significance of MCID and MID in Clinical Research, Practice, and Patient Care

The origin of MCID can be traced back to the seminal works of Guyatt et al. (1987) and Jaeschke et al. (1989), who first introduced the term to quantify the smallest change or difference after a treatment that a patient would perceive as meaningful, perceivable, substantially important, noticeable, or beneficial or even harmful, sufficient to warrant a change in the patient's management (e.g., Cook, 2008; Engel et al., 2018; Eton et al., 2004; Guyatt et al., 2002; Jaeschke et al., 1989; Mokkink et al., 2010a, b; Mouelhi et al., 2020; Revicki et al., 2006, 2008).

In other words, it is about the (patient-)relevant change or the extent of a statistically significant difference or change that the patient perceives as relevant (e.g., Cook, 2008; Copay et al., 2007; Crosby et al., 2003; De Vet et al., 2006, 2007, 2023; Food and Drug Administration, 2009; Jaeschke et al., 1989; Mokkink et al., 2010a, b; Norman et al., 2003, 2004; Revicki et al., 2006, 2008).

The question is whether the extent of a statistically significant difference (between two different groups) or change (i.e., within-participant change or another term "change from baseline") exceeds the established MCID for it to be considered patient-relevant, or, in general, relevant in any field of study, such as biology or the humanities.

Jaeschke et al. (1989) introduced MCID and provides some guidelines for interpreting questionnaire scores. Particularly, they highlighted the significance of incorporating the patient's perspective in determining MCID, which was a new approach at that time. Jaeschke et al. (1989) suggested using PROs and global ratings of change to establish the MCID.

When the extent of a statistically significant difference or change exceeds the established MCID, it suggests that the observed change is not only statistically significant but also clinically meaningful, that is, large enough to have a noticeable impact on the patient, patient care, treatment decisions, or PROs, indicating that the intervention or treatment has a tangible (i.e., patient-relevant) and beneficial effect on patients' health or well-being beyond mere statistical significance.

In sum, the concept of MCID emerged as a response to the limitations of relying solely on statistical significance, which often fails to convey the real-world effectiveness of a treatment (e.g., Jaeschke et al., 1989).

Hence, MCID represents a paradigm shift, emphasizing the importance of the patient's perspective in evaluating treatment efficacy. MCID and MID are concepts that are often used interchangeably. MCID is particularly used in the clinical context to assess whether the effect of a medical intervention has a meaningful, noticeable, and beneficial impact on patient outcomes.

In contrast, MID is a similar but vague term that can be adopted in nonclinical settings. MID reflects the smallest difference in the outcome (e.g., PROs) that patients perceive as beneficial or harmful (Revicki et al., 2000). This value, MID, would change a patient's therapy management or modify their therapy (Revicki et al., 2000). MID is not necessarily confined to clinical interventions (an example for the latter, see, e.g., Bende et al., 2023) in the field of biology or (Fekete et al., 2018) in the domain of psycholinguistics, and can be applied to any change or difference that is considered relevant.

The distinction between these concepts underscores the importance of integrating both clinical judgment and patient perspectives in evaluating the significance of treatment outcomes. MCID focuses on the clinical relevance of changes, guiding therapeutic decisions and reflecting outcomes that alter clinical management (Guyatt et al., 1987, 2002).

On the other hand, MID broadens this scope to include any perceptible improvement valued by patients, highlighting the significance of PROs in clinical research and practice for the indication of anemia and fatigue (Cella et al., 2002). Cella et al. (2002) add that methods for the calculation of MCID should be combined (so-called anchor- and distribution-based methods).

In summary, while both terms aim to quantify the smallest statistically significant change or difference in an outcome that is perceived as important by patients, MCID is more specifically used in clinical research to determine whether a medical intervention has achieved a threshold of clinical significance. MID, on the other hand, has a slightly broader application and can refer to any statistically significant change or difference that has practical relevance. In practice, the distinction between these terms may not always be strictly adhered to, and they are often used interchangeably, which I am going to follow in this book too. The key takeaway is that both MCID and MID focus on the patient's perspective of what is considered an important change, which is critical for assessing the real-world relevance of study findings, medical interventions, and patient care strategies.

The determination of MCID/MID is critical in both clinical trials and practice, as it informs medical decision-making regarding the continuation, alteration, or cessation of a treatment based on its impact on a patient's QoL, symptomatology, or functional status (Jaeschke et al., 1989). It is not merely a statistical significance but rather a clinically meaningful metric, bridging the gap between measurable changes and meaningful PROs (e.g., change in functional status or improvement in respiratory function).

Yet, beneath the surface of this metric lies much uncertainty (e.g., about standardization or the multiplicity of MCIDs/MIDs, to name two), and even the lack of use of the metric across HTA bodies of different countries, as well as within the medical community, let alone outside the clinical field such as in the humanities or social sciences.

MCID serves as a bridge between statistical significance and clinical relevance, ensuring that the results of clinical trials and studies are not only statistically sound but also meaningful to patients (e.g., Revicki et al., 2008). The application of MCID extends beyond clinical trials, influencing clinical decision-making, informing treatment guidelines, and shaping patient-centered care approaches (e.g., Copay et al., 2007).

Integrating MCID/MID into various scientific domains marks a significant advancement in research methodology, enabling the discrimination between statistically significant and practically meaningful results. This book is designed to cater to a broad audience, including health-care practitioners, policymakers, businessmen, and academicians from all fields, with the intent of fostering a comprehensive understanding of MCID/MID and its implications for patient-centered care, health policy, and research in various fields. Through a meticulous examination of MCID's methodologies, applications, and implications, the book contributes to the enhancement of health-care delivery, the formulation of patient-centered policies, and the expansion of research methodologies across all disciplines.

This chapter has established a conceptual understanding of MCID/MID, highlighting its role in bridging statistical and clinical relevance, and its importance in patient-centered care and health care. The subsequent chapters will delve into the methodological approaches of the computation of MCID/MID, the question of variability of multiple metrics, and practical applications of MCID across various clinical and nonclinical settings.

References

Bende, A., László, R., Faragó, S., & Fekete, I. (2023). An investigation of the possibilities for sex and age determination of Eurasian woodcock (*Scolopax rusticola* L.) based on biometric parameters using conditional inference trees and minimal important differences. *Journal of Vertebrate Biology, 73*(23068), 1–15. https://doi.org/10.25225/jvb.23068

Brezina, V., & Meyerhoff, M. (2014). Significant or random? a critical review of sociolinguistic generalisations based on large corpora. *International Journal of Corpus Linguistics, 19*(1), 1–28. https://doi.org/10.1075/ijcl.19.1.01bre

Cella, D., Eton, D. T., Lai, J. S., Peterman, A. H., & Merkel, D. E. (2002). Combining anchor and distribution-based methods to derive minimal clinically important differences on the functional assessment of cancer therapy (FACT) anemia and fatigue scales. *Journal of Pain and Symptom Management, 24*(6), 547–561. https://doi.org/10.1016/s0885-3924(02)00529-8

Cohen, J. (1988a). *Statistical power analysis for the behavioral sciences* (2nd ed.). Lawrence Erlbaum Associates.

Cohen, J. (1988b). The effect size. In *Statistical power analysis for the behavioral sciences* (pp. 77–83). Routledge.

Cook, C. E. (2008). Clinimetrics corner: The minimal clinically important change score (MCID): A necessary pretense. *Journal of Manual & Manipulative Therapy, 16*(4), E82–E83. https://doi.org/10.1179/jmt.2008.16.4.82E

Copay, A. G., Subach, B. R., Glassman, S. D., Polly, D. W., & Schuler, T. C. (2007). Understanding the minimum clinically important difference: A review of concepts and methods. *The Spine Journal, 7*(5), 541–546. https://doi.org/10.1016/j.spinee.2007.01.008

Countrymeters. (2024). India population (2024) live. Retrieved from https://countrymeters.info/en/India

Crosby, R. D., Kolotkin, R. L., & Williams, G. R. (2003). Defining clinically meaningful change in health-related quality of life. *Journal of Clinical Epidemiology, 56*, 395–407. https://doi.org/10.1016/S0895-4356(03)00044-1

De Vet, H. C., Terwee, C. B., Ostelo, R. W., Beckerman, H., Knol, D. L., & Bouter, L. M. (2006). Minimal changes in health status questionnaires: Distinction between minimally detectable change and minimally important change. *Health and Quality of Life Outcomes, 4*, 54. https://doi.org/10.1186/1477-7525-4-54

De Vet, H. C. W., Ostelo, R. W., Terwee, C. B., Van der Roer, N., Knol, D. L., Beckerman, H., et al. (2007). Minimal important change determined by a visual method integrating an anchor-based and a distribution-based approach. *Quality of Life Research, 16*(1), 131–142. https://doi.org/10.1007/s11136-006-9109-9

De Vet, H. C. W., Terwee, C. B., & Mokkink, L. B. (2023). Minimal important difference. In F. Maggino (Ed.), *Encyclopedia of quality of life and well-being research*. Springer. https://doi.org/10.1007/978-3-031-17299-1_1813

Engel, L., Beaton, D. E., & Touma, Z. (2018). Minimal clinically important difference: A review of outcome measure score interpretation. *Rheumatic Disease Clinics of North America, 44*(2), 177–188. https://doi.org/10.1016/j.rdc.2018.01.011

Engsted, T. (2009). Statistical vs. economic significance in economics and econometrics: Further comments on McCloskey and Ziliak. *Journal of Economic Methodology, 16*(4), 393–408. https://doi.org/10.1080/13501780903337339

Eton, D. T., Cella, D., Yost, K. J., Yount, S. E., Peterman, A. H., Neuberg, D. S., Sledge, G. W., & Wood, W. C. (2004). A combination of distribution- and anchor based approaches determined minimally important differences (MIDs) for four endpoints in a breast cancer scale. *Journal of Clinical Epidemiology, 57*, 898–910. https://doi.org/10.1016/j.jclinepi.2004.01.012

Fekete, I., Schulz, P., & Ruigendijk, E. (2018). Exhaustivity in single bare wh-questions: A differential-analysis of exhaustivity. *Glossa: A Journal of General Linguistics, 3*(1), 96. https://doi.org/10.5334/gjgl.549

Field, A. (2009). *Discovering statistics using SPSS* (3rd ed.). Sage.

Fisher, R. A. (1925). *Statistical methods for research workers*. Oliver and Boyd.

Food and Drug Administration. (2009, December). *Guidance for industry: Patient-reported outcome measures: Use in medical product development to support labeling claims*. U.S. Department of Health and Human Services.

Guttikunda, S. K., & Goel, R. (2013). Health impacts of particulate pollution in a megacity—Delhi, India. *Environmental Development, 6*, 8–20. https://doi.org/10.1016/j.envdev.2012.12.002

Guyatt, G., Walter, S., & Norman, G. (1987). Measuring change over time: Assessing the usefulness of evaluative instruments. *Journal of Chronic Diseases, 40*(2), 171–178. https://doi.org/10.1016/0021-9681(87)90069-5

Guyatt, G. H., Osoba, D., Wu, A. W., Wyrwich, K. W., & Norman, G. R. (2002). Methods to explain the clinical significance of health status measures. *Mayo Clinic Proceedings, 77*(4), 371–383. https://doi.org/10.4065/77.4.371

Jacobson, N. S., Roberts, L. J., Berns, S. B., & McGlinchey, J. B. (1999). Methods for defining and determining the clinical significance of treatment effects: Description, application, and alternatives. *Journal of Consulting and Clinical Psychology, 67*(3), 300–307. https://doi.org/10.1037/0022-006X.67.3.300

Jaeschke, R., Singer, J., & Guyatt, G. H. (1989). Measurement of health status: Ascertaining the minimal clinically important difference. *Controlled Clinical Trials, 10*(4), 407–415. https://doi.org/10.1016/0197-2456(89)90005-6

Kazdin, A. E. (1999). The meanings and measurement of clinical significance. *Journal of Consulting and Clinical Psychology, 67*(3), 332–339. https://doi.org/10.1037/0022-006X.67.3.332

King, M. T. (2011). A point of minimal important difference (MID): A critique of terminology and methods. *Expert Review of Pharmacoeconomics & Outcomes Research, 11*(2), 171–184. https://doi.org/10.1586/erp.11.9

Kline, R. B. (2013). *Beyond significance testing: Statistics reform in behavioral sciences* (2nd ed.). American Psychological Association. https://doi.org/10.1037/14136-000

Koplenig, A. (2019). Against statistical significance testing in corpus linguistics. *Corpus Linguistics and Linguistic Theory, 15*(2), 321–346. https://doi.org/10.1515/cllt-2016-0036

Kumar, P., Gulia, S., Harrison, R. M., & Khare, M. (2017). The influence of odd-even car trial on fine and coarse particles in Delhi. *Environmental Pollution, 20–30*, 225. https://doi.org/10.1016/j.envpol.2017.03.017

Lin, M., Lucas, H. C., Jr., & Shmueli, G. (2013). Too big to fail: Large samples and the p-value problem. *Information Systems Research, 24*(4), 906–917. https://doi.org/10.1287/isre.2013.0480

Mokkink, L. B., Terwee, C. B., Patrick, D. L., Alonso, J., Stratford, P. W., Knol, D. L., Bouter, L. M., & de Vet, H. C. (2010a). The COSMIN checklist for assessing the methodological quality of studies on measurement properties of health nature measurement instruments: An international Delphi study. *Quality of Life Research, 19*, 539–549. https://doi.org/10.1007/s11136-010-9606-8

Mokkink, L. B., Terwee, C. B., Patrick, D. L., Alonso, J., Stratford, P. W., Knol, D. L., Bouter, L. M., & de Vet, H. C. (2010b). The COSMIN study reached international consensus on taxonomy, terminology, and definitions of measurement properties for health-related patient-reported outcomes. *Journal of Clinical Epidemiology, 63*(7), 737–745. https://doi.org/10.1016/j.jclinepi.2010.02.006

Mouelhi, Y., Jouve, E., Castelli, C., & Gentile, S. (2020). How is the minimal clinically important difference established in health-related quality of life instruments? Review of anchors and methods. *Health and Quality of Life Outcomes, 18*(1), 136. https://doi.org/10.1186/s12955-020-01344-w

Nagpure, A. S., Gurjar, B. R., & Martel, J. C. (2014). Human health risks in national capital territory of Delhi due to air pollution. *Atmospheric Pollution Research, 5*(3), 371–380. https://doi.org/10.5094/APR.2014.043

Norman, G. R., Sloan, J. A., & Wyrwich, K. W. (2003). Interpretation of changes in health-related quality of life: The remarkable universality of half a standard deviation. *Medical Care, 41*(5), 582–592. https://doi.org/10.1097/01.MLR.0000062554.74615.4C

Norman, G. R., Sloan, J. A., & Wyrwich, K. W. (2004). The truly remarkable universality of half a standard deviation: Confirmation through another look. *Expert Review of Pharmacoeconomics & Outcomes Research, 4*(5), 581–585. https://doi.org/10.1586/14737167.4.5.581

Revicki, D., Osoba, D., Fairclough, D., et al. (2000). Recommendations on health-related quality of life research to support labeling and promotional claims in the United States. *Quality of Life Research, 9*, 887–900. https://doi.org/10.1023/A:1008996223999

Revicki, D. A., Cella, D., Hays, R. D., Sloan, J. A., Lenderking, W. R., & Neil, K. A. (2006). Responsiveness and minimal important differences for patient reported outcomes. *Health and Quality of Life Outcomes, 4*(70), 1–5. https://doi.org/10.1186/1477-7525-4-70

Revicki, D. A., Hays, R. D., Cella, D., & Sloan, J. (2008). Recommended methods for determining responsiveness and minimally important differences for patient-reported outcomes. *Journal of Clinical Epidemiology, 61*(2), 102–109. https://doi.org/10.1016/j.jclinepi.2007.03.012

Sullivan, G. M., & Feinn, R. (2012). Using effect size-or why the P value is not enough. *Journal of Graduate Medical Education, 4*(3), 279–282. https://doi.org/10.4300/JGME-D-12-00156.1

Wasserstein, R. L., & Lazar, N. A. (2016). The ASA'S statement on p-values: Context, process, and purpose. *The American Statistician, 70*(2), 129–133. https://doi.org/10.1080/00031305.2016.1154108

Wasserstein, R. L., Schirm, A. L., & Lazar, N. A. (2019). Moving to a world beyond "$p < 0.05$". *The American Statistician, 73*(sup1), 1–19. https://doi.org/10.1080/00031305.2019.1583913

Worldometer. (2024). India Population (2024) – Worldometer. Retrieved from https://www.worldometers.info/world-population/india-population/

Chapter 2
Theoretical Foundations of MCID and MID: Put the Patient or Client at the Forefront

Abstract Differentiating between statistical significance and practical significance—commonly referred to as clinical relevance, meaningful difference, or minimal clinically important/important difference (MCID/MID)—is crucial for sound evaluation within a scientific framework. While *p*-values provide a probabilistic assessment of data within the scope of the null hypothesis, they remain silent on the matter of whether the associated effect holds any significance in real-world scenarios. Some researchers critique the prevalent misuse of *p*-values, arguing that practical or clinical significance is often overshadowed by the focus given to statistical significance. Such misinterpretation is bound to result in erroneous choices and actions in health, policy, or business. Instead, Bayesian statistics are offered as a remedy, albeit they do not resolve the issue of defining practical relevance using MCID/MID thresholds.

Statistical significance, specifically the *p*-value threshold of $p < 0.05$, has received critique from various corners for being arbitrary. With sufficiently large sample sizes, even the most trivial of effects will be deemed to be statistically significant. Hence, to avoid this theoretical trap, effect sizes and confidence intervals, along with relevant domain expertise and patient-reported outcomes (PROs), need to be integrated. Clinical relevance goes a step further and ascertains whether the treatment effect is perceivable, justifiable, and whether the benefits outweigh the costs of intervention.

MCID, which applies primarily to health care, also informs other domains such as linguistics and biology. It is crucial to note that MCID is not ubiquitous. The chapter demonstrates how employing different approaches contributes to considerable variability in MCID estimates, underscoring the need for customized or stratified methods. In the end, the chapter strongly advocates for the blend of MCID/MID with what is called “practical relevance” to support statistical outcomes that are not only significant but also meaningful and useful in real-world applications.

Keywords Clinical significance · Statistical significance · Effect size · Patient-centered outcomes · PROs · Clinical importance

I. Fekete, *Transforming Data into Informed Decisions across Clinical and Non-Clinical Domains*, https://doi.org/10.1007/978-3-032-02595-1_2

The Difference Between Clinical Significance vs Statistical Significance

Let us now return to the difference between *statistical significance* and *practical significance* (or "clinical relevance," "clinical significance," "practical importance," "meaningful difference," "substantial difference," "meaningful change," etc.). Note that these terms are usually used interchangeably, unless specified otherwise.

Greenland et al. (2016) have critically examined common misunderstandings surrounding statistical tests and *p*-values by explicitly addressing the issue of mistaking *statistical significance* for something else. It is worthwhile reading the article because it provides 25 misinterpretations of *p*-values, confidence intervals (CIs), and power (Greenland et al., 2016).

What is important for us at this point is that, for example, Kline (2013) and Greenland et al. (2016) both emphasize that statistical significance is frequently *misinterpreted* or even *overstated* (!) in its implications for research findings. As a solution, Kline (2013) advocates Bayesian statistics in psychology as an alternative to traditional hypothesis testing. However, even Bayesian statistics would not solve our problem of determining the threshold of MID/MCID and the concept of practical implication. In my view, it just solves the problem of *p*-value inflation and its reliability, which is a great step forward.

Because the threshold of a *p*-value of less than 0.05 is *arbitrary* and has been subject to much criticism (e.g., Kline, 2013; Wasserstein et al., 2019), Wasserstein et al. (2019) propose the use of Bayesian statistics, meta-analyses, and evidence reports.

Misinterpretations and overstatements arise when researchers, or more often the general public and the media, equate statistical significance with practical or clinical significance. This phenomenon neglects real-world applications and patient care, a problem that MCID/MID can combat.

Crucially, *statistical significance* does not imply *practical relevance* (Jaeschke et al., 1989). To put it simply, the effect observed is not large or important enough to act or make an informed and enhanced decision (any type of action; in health care, politics, business, economics, etc.) unless it reaches practical relevance (i.e., the MID).

Another concern is that researchers might unconsciously or deliberately manipulate the data to achieve statistically significant results, for example, by increasing the sample size arbitrarily (*p*-hacking). I am going to show a few studies later in the book, specifically in which statistical significance was achieved through sample-size manipulation (i.e., the increase of sample-size until significance is achieved).

Greenland et al. (2016) claim that statistical significance should be viewed as a probabilistic statement about data given a specific statistical model, and it requires careful interpretation within the context of study design, effect sizes, confidence intervals (CIs), and the broader research domain. For Greenland et al. (2016), the critical take-home message is that statistical significance is a tool for inference, not a definitive proof of a hypothesis. Hence, it must be integrated with other statistical measures and domain-specific knowledge (sometimes called "subject-matter knowledge") such as expert opinions (expert panels, Delphi method), for example, by medical professionals, to draw meaningful conclusions from data.

Clinical relevance, on the other hand, considers whether the magnitude of a statistically significant difference or change is meaningful (from the perspective of the patient or the MD), important, perceivable, or noticeable in terms of PROs, public health impact, resource allocations, or policy implications (e.g., Copay et al., 2007; Crosby et al., 2003; Engel et al., 2018; Eton et al., 2004; Farivar et al., 2004; FDA, 2009; Guyatt et al., 1987, 2002; Mokkink et al., 2010a, b; Norman et al., 2003; Revicki et al., 2006, 2008).

In the domain of air pollution interventions in India, a clinically significant outcome would require not only statistical significance but also a considerable and meaningful *enhancement in air quality* and public health indicators, PROs (such as a *notable decrease in hospitalizations* or *a notable decrease in respiratory incidents*), demonstrating a tangible advantage to the population's welfare, health-care system, and infrastructure.

It would be fairly easy to conduct an experiment on nonsmokers using a longitudinal study with multiple measurements. The clinical endpoint would be the different measures of pulmonary function. Patients have to inhale and exhale into a tube forcefully. Any intervention (reduction in AQI) by any means (lockdown, odd–even vehicle scheme, introduction of electric cars, bioethanol, etc.) should lead to a substantial improvement in pulmonary function. But how far should we go in reduction in points on the AQI scale? The test to assess pulmonary function is called the pulmonary function test (PFT). Once we reach the point of substantial improvement on the PFT, we can compute the reduction of AQI: how many points of reduction are required to improve the pulmonary function of the patient. And we will get the MCID in points.

If you reread the previous paragraph, you can realize that patient characteristics—except for gender and age—are not included in this hypothetical experiment. Even ethnicity might count too. However, there is now an ongoing debate over the inclusion of the so-called "race" (sic) in the PFT. (Instead of "race," I would use ethnicity because every human belongs to one race.) The American Thoracic Society (ATS) now advocates a so-called "race-neutral PFT."

Anyway, I would just like to stress how the MCID of AQI can be computed using a longitudinal study. In addition, I advocate at least a stratification by gender and age for the computation of MCID, as these two variables influence pulmonary function. As a result, we will get four strata, i.e., four MCIDs.

For example, a reduction in respiratory and cardiovascular diseases among the population or a significant improvement in AQI leading to fewer COPD cases, fewer hospital admissions for respiratory issues, etc., could serve as indicators of practical relevance of the reduction in AQI. If MDs also confirm that they have perceivably less work and fewer air-quality-related admissions or a noticeable reduction in COPD cases, for example, then that can serve as an informal indicator of the reduction having exceeded the MCID for the given disease due to some intervention or policy (personal communication with a pulmonologist).

Let me give another informal example instead which clearly elucidates MCID. Almost everyone knows that on New Year's Eve, hospitals are overburdened due to drug or alcohol intoxication contrary to the rest of the days of the year. In other words, the number of patients even exceeds statistical significance, maybe even the MCID. Toxicologists report almost every year that they have a lot of work on New Year's Eve, which strongly suggests that hospitalization as a clinical endpoint has exceeded the MCID of the number of toxicology patients. The opposite is also true. According to anecdotal evidence, some MDs have observed that patients do not come on holidays despite their free time and their condition (personal communication with an MD).

Regarding AQI, a similar dilemma is encountered with its scale ranging from 0 to 500.[1] A reduction of 50 points or even more is required to transition from one category to another on the six-category scale, as the scale is segmented into six categories (US Environmental Protection Agency, 2014; EPA). An important question arises as to whether a decrease of 50 points worth a lot of money if no noticeable change can be observed in diseases.

To underscore, I have not calculated the MCID for this scale because I did not have an external clinical endpoint to take it as the "anchor" against which I can perform the MCID calculation. However, I wish to highlight the necessity of establishing an MCID for the AQI as well because any relevant change in AQI that leads to shifting from one category to another can be considered important for public health.

Determining a precise value for the MCID for AQI necessitates comprehensive epidemiological research to correlate particular alterations in AQI levels with clinically significant health consequences (PROs) or enhancements in public health. To put it simply, if we just want to decrease visibility (on the road or at airports), the MID of AQI would be different from health-related MID such as the decrease of COPD hospitalizations.

In other words, a transition from one category to another on the AQI scale might be statistically significant but might have zero effect on the health status of patients. Moreover, these categories are arbitrary. On November 4, 2024, a very high record was reported[2] in Lahore, Pakistan; the AQI value reached 1900 (sic), even leading to a lockdown. This AQI value is nearly 40 times the limit of the WHO that is considered safe.

As you might have noticed, we have shifted from the hypothetical air pollution study to AQI, and we can detect MCID "everywhere"; I am convinced of its universal relevance and importance. It is crucial to clarify the specific outcomes for which the MCID/MID is intended to be calculated: are we focusing on long-term or short-term respiratory health outcomes, the number of new COPD cases, the number of

---

[1] https://www.aqi.in/

[2] https://www.thehealthsite.com/news/lockdown-imposed-complete-shutdown-of-schools-and-colleges-declared-as-pakistan-aqi-breaches-1900-mark-lahore-news-1147360/

relapses, the frequency of hospital visits, the length of hospitalization, or some ecological indicators, among others?

It is important to note that any reduction in AQI must be evaluated *against a well-defined threshold or variable*. The choice of this reference point is very important, as the MCID/MID is likely to vary based on the outcome (e.g., the PRO) or variable measured (e.g., the visibility). Later in the book, I am going to enlist a bunch of other patient-related and context-dependent variables that affect the MCID. I am going to show that it is possible that the MCID/MID, being a dynamic concept, changes within a person too. Just think of individualized medicine. The dosage of a drug is contingent on gender, age, etc. We will talk about this topic later on, and I am going to prove that the MCID for the elderly for certain drugs is different due to their condition.

Another argument against standardization is that, for instance, it can be noted that attaining long-term health benefits requires a significantly larger decrease in the AQI (e.g., 300 points) than the drop needed for ecological benefits, such as enhancements in vegetation health and biodiversity (e.g., 180 points).

MCID should be calculated in relation to the outcomes, clinical endpoints, including respiratory events, cardiovascular illnesses, frequency of hospital admissions, duration of hospitalization, or patient count. Without a definitive MCID, the daily variations and reductions in AQI may be misleading, creating an illusion of improvement in health.

MCID represents an important metric, with a practical impact of medical interventions on patients' lives and well-being (e.g., Cook, 2008; Engel et al., 2018; Eton et al., 2004; Jaeschke et al., 1989; Mokkink et al., 2010a, b; Revicki et al., 2006, 2008; Terwee et al., 2009).

This book is designed with several goals in mind: firstly, to familiarize the reader with the concepts of MCID and MID (and similar metrics); secondly, to showcase the utility of these metrics in research and practice; and thirdly, to advocate the application of these metrics in nonclinical fields, including biology (e.g., Bende et al., 2023), linguistics (e.g., Fekete et al., 2018), the social sciences, environmental sciences, the humanities, or even the business industry.

I am going to elaborate on the concept of "practical relevance" of a statistically significant difference or change analogous to the concept of MCID/MID. "Practical relevance" refers to the extent to which a statistical finding is meaningful, useful, noticeable, impactful, relevant, practical or substantial in real-world applications, beyond "just" being statistically significant (e.g., Engel et al., 2018; Eton et al., 2004; Guyatt et al., 1987, 2002; Jaeschke et al., 1989; Mokkink et al., 2010a, b; Revicki et al., 2006, 2008; Terwee et al., 2009). I strived to come up with as many adjectives as possible to capture the fundamental nature of MCID.

For instance, in biology, an established indicator such as "biologically relevant difference" would denote a statistically significant difference which is relevant to a biologist and in biology, or, in other words, which reflects a meaningful difference or change in a biological study.

In the business world, it would refer to a difference or change that requires some intervention or immediate reaction (e.g., "sell" or "buy"). Crucially, statistical significance is neither a necessary nor sufficient criterion for determining "biological relevance" (Nakagawa & Cuthill, 2007).

This book is also for professionals outside the health-care industry, such as the business world, linguistics, members of NGOs who gather survey data, the social sciences, or biology, among others, as MID can be utilized in almost every scientific discipline where traditional hypothesis testing is performed. This book is intended for health-care practitioners, clinicians, physicians, other health-care providers, clinical researchers, health policymakers, students in health-care disciplines (medical, nursing, and allied health students as well as those studying public health, biostatistics, and health services research), academics and educators in medical and health science programs, health-care quality improvement professionals, but also for health-care informatics and data scientists, Health Economics and Outcomes Research (HEOR) professionals, and regulatory affairs professionals. But, most importantly, it is intended to those researchers in academia as well as in the industry who are outside health care.

For example, individuals from the business industry may find the insights about MID offered in this book to be of substantial benefit for their company. Company data corroborated with MID can provide businesses with a competitive edge. It can help organizations to understand both their customers and employees better, optimize their operations, and make more informed and optimal strategic decisions (instead of, e.g., *under- or overreacting to statistically significant but trivial trends*).

In the business world, the application of MIDs is important across various sectors such as company records, the productivity of employees as measured by questionnaires, for example, the success rate of a marketing campaign or that of a new policy implemented in a company.

MIDs provide the opportunity to *improve consumer experience*. Businesses can even tell their customers about MID to *gain leverage* and showcase their *scientific professionalism*. For example, they can *mention MID on flyers*. By analyzing customer behavior via data, firms may customize their offerings and products, predict customer demands, and cultivate closer relationships with their clients. This may result in *enhanced consumer loyalty* and, hence, *elevated profitability*.

With an MID as an extra statistical tool, businesses could also *maintain the trust of their customers and stakeholders, forecast future trends and behaviors, enabling businesses to proactively respond to changes*. MIDs can also enhance customer experience by providing personalized and targeted services, thereby increasing not just customer loyalty but also customer retention. MIDs can both uncover nontrivial trends and, at the same time, filter out trivial ones that might otherwise go unnoticed. This "filtering process" is the heart of informed and strategic decision-making, the novelty of the book.

Businesses frequently employ customer satisfaction surveys to assess consumer sentiment. An MID in this context indicates the minimal change in survey scores that bears significance for customers. In the absence of an MID, enterprises may misallocate resources or overlook chances to improve customer experience. Moreover, companies can detect inefficiencies and wasteful practices, resulting in cost reductions and improved productivity. I am going to devote an extra chapter to the business world, as I deem it important when we talk about MID.

This book propagates the incorporation of MCID/MID in guidelines, policy frameworks, clinical and psychological research, statistical analysis plans (SAPs), preregistration reports, clinical trials for early benefit assessment, and in any study or industry report also outside nonclinical fields. This book presents a viewpoint on evaluating treatment efficacy, promoting a transition toward outcomes that are significantly advantageous to patients. Through a critical analysis of the techniques and applications of MCID/MID, the book would substantially enhance patient-centered care, the broader domain of health outcomes research, and several disciplines beyond medicine and psychology.

A change is occurring in patient management, shifting from a conventional, physician-centric, biomedical model to a more patient-centered and individualized approach (Institute of Medicine, 2001). This new paradigm emphasizes the patient's personal experience (PROs) and their particular requirements. This is the juncture at which data transforms into improved, patient-centered decisions capable of revolutionizing health care. Patient-centered care and shared decision making (SDM) beyond mere disease treatment involves addressing the individual, acknowledging their distinctiveness, comprehending their specific requirements, and customizing care plans accordingly (Barry & Edgman-Levitan, 2012).

The *book will propose a new concept of "practical relevance" for nonclinical fields* analogous to the concept of MCID in the clinical field (i.e., MID). Such a concept or metric is not yet well established in nonclinical fields, which unfortunately results in the acceptance of statistically significant but irrelevant/practically not meaningful results, differences or changes.

Let us go one step backward to statistical significance and put aside MCID/MID for a moment. Addressing the notion of statistical significance, it becomes evident that political actors might portray and communicate the results of this hypothetical air pollution study as a successful achievement without commenting on the *practical relevance of the reduction* (i.e., MID).

They could do it not necessarily through direct manipulation, fabrication, or falsification of the data or its statistical analysis. Instead, they may simply disregard, neglect, or be unaware of the concept of MCID and exploit the statistical naivety and ignorance of the general population. However, this book does not delve into this particularly sensitive issue of misuse of statistics. After all, we do not mean to hand out "cheat sheets" that might tempt policymakers, politicians, or researchers to gain leverage through unfair means.

The distinction between *clinical significance* and *statistical significance* is crucial in the interpretation of PROs (Jaeschke et al., 1989; Revicki et al., 2008). Statistical significance (Fisher, 1925) and confidence intervals (CIs) are not enough to make informed and reliable decisions (Cumming, 2014). The problem is that the *p*-value does not tell us anything about the magnitude, importance, or extent of these differences in real-world scenarios.

This is where the concept of clinical significance comes into the picture (e.g., Cook, 2008; Copay et al., 2007; Crosby et al., 2003; De Vet et al., 2006, 2007, 2023; Engel et al., 2018; Farivar et al., 2004; FDA, 2009; Jaeschke et al., 1989; Mokkink et al., 2010a, b; Norman et al., 2003; Revicki et al., 2006, 2008).

The Relationship Between Effect Size, Clinical Importance, and Patient-Centered Outcomes

Clinical relevance is a broader and more inclusive concept, considering the size of the effect (e.g., Cohen, 1988), the costs and benefits (see, e.g., HTA authorities and health insurance companies), the PROs, the real-world applicability of the findings ("practical relevance"), and the value judgment of the stakeholders. Clinical relevance evaluates whether the statistical difference in outcomes has a meaningful impact on real-world decisions and patient care, and it is a measure of the practical importance of a treatment effect.

Crucially, a large sample size can detect a statistically significant difference, even when the actual difference is *negligible/trivial* in real-world terms (e.g., Angst et al., 2017; Brezina & Meyerhoff, 2014; Lin et al., 2013). In sum, a study may produce results that are *statistically significant*, but *not clinically relevant*. For instance, a drug may statistically reduce the incidence of a disease, but the reduction may be so small that it does not justify the cost, side effects, or other implications of the treatment. Another tangential topic is that other contributing factors should be ruled out to prove causality.

An instance where a drug failed to achieve the MCID is Neumora Therapeutics' experimental major depression medication under the composition name *navacaprant* (January 2, 2025).[3] In the Phase 3 clinical trial, navacaprant did not achieve justifiable efficacy when compared to a placebo in reductions of depressive symptoms.[4] No significant clinical improvements were observed in the drug's treatment group, as well as the placebo group, which both averaged a 12.5 point decrease in the Montgomery-Åsberg Depression Rating Scale (MADRS) score. It is also unclear to me why they took placebo as the control group because it should actually be best practice (SoC; standard of care) as per IQWiG guidelines (IQWiG, 2023). In other words, the safety profile and efficacy of the new drug should be compared to the SoC; it should surpass the control group (SoC/best practice).

MCID, introduced one year after Cohen (1988), by Jaeschke et al. (1989), is mainly relevant in clinical settings where the determination of whether a treatment effect is large enough to warrant a change in practice is critical. It surpasses

[3] https://ir.neumoratx.com/news-releases/news-release-details/neumora-therapeutics-reports-data-koastal-1-study-navacaprant#:~:text=About%20Navacaprant&text=Navacaprant%20is%20an%20investigational%20once,cognition%2C%20reward%2C%20and%20behavior

[4] https://www.barrons.com/articles/neumora-stock-depression-drug-fails-10b8ed28?utm_source=chatgpt.com

statistical significance and considers the smallest benefit of clinical interest that would justify the use or change in a patient's treatment plan (Jaeschke et al., 1989).

But when it comes to HTA, the new drug should exceed the best practice. In the previous example, best practice (also known as SoC; standard of care) could be another patient group with the same indication and exclusion criteria. Importantly, each and every participant in the study would receive best-practice treatment adjusted to their needs, demographics, and clinical status. In other words, the experimental group and the control group should be comparable on the baseline characteristics and should have the same clinical indication (in the previous example: depression).

Let us talk about the relationship between effect size and MCID/MID. Cohen (1988) delineated effect sizes into three ranges, providing a framework for researchers to assess the strength of relationships or differences in their data. However, the necessity of MCID, MID, and Cohen's effect size since its introduction in 1988 (Cohen, 1988) all provide a comprehensive evaluation of research outcomes. Their integrated approach, for instance, as in Fekete et al. (2018), enriches the interpretation of clinical and psychological research among other fields and also gives supporting evidence to research findings.

A study may report a statistically significant effect classed as "large" or "medium" by Cohen's (1988) guidelines, although this effect could still be below the thresholds set for MCID or MID. This scenario may occur in various settings. For example, measurement precision and scale sensitivity may influence the study instruments or scales employed, which could be highly sensitive and capable of detecting minute changes that, although statistically significant, lack therapeutic or practical relevance.

I seek to elucidate the difference between effect size measures and MCID/MID. Franceschini et al. (2023) conducted a cohort study, classified as level 3 evidence, examining the discrepancies in MCID estimations for individuals with knee osteoarthritis (OA). The "International Knee Documentation Committee" (IKDC) subjective score at 6 months was applied to a dataset consisting of 312 patients in order to determine the MCIDs, which performed by applying 17 different MCID approaches simultaneously (Franceschini et al., 2023).

The MCID values showed great variability, resulting in substantial discrepancies in the percentage of patients with MCID values ranging from 24.0% to 66.0% for "anchor-based methods" and 44.6% to 75.9% for "distribution-based methods" (these two methodologies will be elaborated upon later in the book).

Franceschini et al. (2023) illustrate that while effect sizes are stable, there is considerable variability in MCID values, an argument again against a universal MCID, in my opinion. Obviously, stratification would explain this variability in MCIDs, but such a great variability indicates the need for incorporating demographic and clinical variables in a model to *personalize the MCID*, in my opinion.

Another compelling systematic review and meta-analysis that exemplifies the distinction between effect size and MCID is the one conducted by Tubach et al. (2012) from the field of rheumatology. Tubach et al. (2012) identify the MID thresholds for both improvement and worsening across PROs in rheumatology. Tubach et al. (2012) demonstrate how these two measures can yield different results about the effectiveness of medical interventions. Essentially, there is one way of calculating, for instance, Cohen's effect size (*d*), but there are multiple methods of doing so for MCID/MID. We are "left in the dark" as to which method to choose to compute MCID/MID.

One of the main questions of this book of whether a threshold analogous to MCID, termed, for example, "practical relevance," could be established also in the social sciences, among other disciplines. In social sciences, the notion of "practical relevance" functions to connect statistical significance (i.e., an indicator of the likelihood that discoveries are not attributable to chance) with the real-world implications of these findings.

Linguistics, for example, may significantly benefit from such a criterion. It would assess the practical significance of differences and changes regarding their relevance. An instance of the application of MID in linguistics is the research conducted by Fekete et al. (2018). A further illustration from the domain of biology is Bende et al. (2023). These two studies highlight the necessity for a metric such as "practical relevance," as they *revealed several statistically significant results that are below a specified MCID-like threshold (MID).*

In sum, by employing MCID and MID, we can prioritize PROs, the patient's subjective assessment of treatment effects, QoL, and functional improvement. Crucially, it must be ensured that clinical decision-making aligns with what is most beneficial to the patient (FDA, 2009).

References

Angst, F., Aeschlimann, A., & Angst, J. (2017). The minimal clinically important difference (MCID) raised the significance of outcome effects above the statistical level, with methodological implications for future studies. *Journal of Clinical Epidemiology, 82*, 128–136. https://doi.org/10.1016/j.jclinepi.2016.11.016

Barry, M. J., & Edgman-Levitan, S. (2012). Shared decision making--pinnacle of patient-centered care. *The New England Journal of Medicine, 366*(9), 780–781. https://doi.org/10.1056/NEJMp1109283

Bende, A., László, R., Faragó, S., & Fekete, I. (2023). An investigation of the possibilities for sex and age determination of Eurasian woodcock (*Scolopax rusticola* L.) based on biometric parameters using conditional inference trees and minimal important differences. *Journal of Vertebrate Biology, 73*(23068), 1–15. https://doi.org/10.25225/jvb.23068

Brezina, V., & Meyerhoff, M. (2014). Significant or random?: A critical review of sociolinguistic generalisations based on large corpora. *International Journal of Corpus Linguistics, 19*(1), 1–28. https://doi.org/10.1075/ijcl.19.1.01bre

Cohen, J. (1988). *Statistical power analysis for the behavioral sciences* (2nd ed.). Lawrence Erlbaum Associates. https://doi.org/10.4324/9780203771587

Cook, C. E. (2008). Clinimetrics corner: The minimal clinically important change score (MCID): A necessary pretense. *Journal of Manual & Manipulative Therapy, 16*(4), E82–E83. https://doi.org/10.1179/jmt.2008.16.4.82E

Copay, A. G., Subach, B. R., Glassman, S. D., Polly, D. W., & Schuler, T. C. (2007). Understanding the minimum clinically important difference: A review of concepts and methods. *The Spine Journal, 7*(5), 541–546. https://doi.org/10.1016/j.spinee.2007.01.008

Crosby, R. D., Kolotkin, R. L., & Williams, G. R. (2003). Defining clinically meaningful change in health-related quality of life. *Journal of Clinical Epidemiology, 56*, 395–407. https://doi.org/10.1016/S0895-4356(03)00044-1

Cumming, G. (2014). The new statistics: Why and how. *Psychological Science, 25*(1), 7–29. https://doi.org/10.1177/0956797613504966

De Vet, H. C., Terwee, C. B., Ostelo, R. W., Beckerman, H., Knol, D. L., & Bouter, L. M. (2006). Minimal changes in health status questionnaires: Distinction between minimally detectable change and minimally important change. *Health and Quality of Life Outcomes, 4*, 54. https://doi.org/10.1186/1477-7525-4-54

De Vet, H. C. W., Ostelo, R. W., Terwee, C. B., Van der Roer, N., Knol, D. L., Beckerman, H., et al. (2007). Minimal important change determined by a visual method integrating an anchor-based and a distribution-based approach. *Quality of Life Research, 16*(1), 131–142. https://doi.org/10.1007/s11136-006-9109-9

De Vet, H. C. W., Terwee, C. B., & Mokkink, L. B. (2023). Minimal important difference. In F. Maggino (Ed.), *Encyclopedia of quality of life and Well-being research*. Springer. https://doi.org/10.1007/978-3-031-17299-1_1813

Engel, L., Beaton, D. E., & Touma, Z. (2018). Minimal clinically important difference: A review of outcome measure score interpretation. *Rheumatic Disease Clinics of North America, 44*(2), 177–188. https://doi.org/10.1016/j.rdc.2018.01.011

Eton, D. T., Cella, D., Yost, K. J., Yount, S. E., Peterman, A. H., Neuberg, D. S., Sledge, G. W., & Wood, W. C. (2004). A combination of distribution- and anchor based approaches determined minimally important differences (MIDs) for four endpoints in a breast cancer scale. *Journal of Clinical Epidemiology, 57*, 898–910. https://doi.org/10.1016/j.jclinepi.2004.01.012

Farivar, S. S., Liu, H., & Hays, R. D. (2004). Half standard deviation estimate of the minimally important difference in HRQOL scores? *Expert Review of Pharmacoeconomics & Outcomes Research, 4*(5), 515–523. https://doi.org/10.1586/14737167.4.5.515

FDA. (2009). *Guidance for industry: Patient-reported outcome measures: Use in medical product development to support labeling claims*. Food and Drug Administration.

Fekete, I., Schulz, P., & Ruigendijk, E. (2018). Exhaustivity in single bare wh-questions: A differential-analysis of exhaustivity. *Glossa: A Journal of General Linguistics, 3*(1), 96. https://doi.org/10.5334/gjgl.549

Fisher, R.A. (1925). *Statistical methods for research workers* (11th ed. rev.). Oliver and Boyd: Edinburgh.

Franceschini, M., Boffa, A., Pignotti, E., Andriolo, L., Zaffagnini, S., & Filardo, G. (2023). The minimal clinically important difference changes greatly based on the different calculation methods. *The American Journal of Sports Medicine, 51*(4), 1067–1073. https://doi.org/10.1177/03635465231152484

Greenland, S., Senn, S. J., Rothman, K. J., Carlin, J. B., Poole, C., Goodman, S. N., & Altman, D. G. (2016). Statistical tests, P values, confidence intervals, and power: A guide to misinterpretations. *European Journal of Epidemiology, 31*(4), 337–350. https://doi.org/10.1007/s10654-016-0149-3

Guyatt, G., Walter, S., & Norman, G. (1987). Measuring change over time: Assessing the usefulness of evaluative instruments. *Journal of Chronic Diseases, 40*(2), 171–178. https://doi.org/10.1016/0021-9681(87)90069-5

Guyatt, G. H., Osoba, D., Wu, A. W., Wyrwich, K. W., & Norman, G. R. (2002). Methods to explain the clinical significance of health status measures. *Mayo Clinic Proceedings, 77*(4), 371–383. https://doi.org/10.4065/77.4.371

Institute of Medicine (IOM) (2001). *Crossing the quality chasm: A New health system for the 21st Century*. Washington, DC: The National Academies Press.

IQWiG. (2023). *General methods version 7.0*, https://www.iqwig.de/methoden/general-methods_version-7-0.pdf
Jaeschke, R., Singer, J., & Guyatt, G. H. (1989). Measurement of health status: Ascertaining the minimal clinically important difference. *Controlled Clinical Trials, 10*(4), 407–415. https://doi.org/10.1016/0197-2456(89)90005-6
Kline, R. B. (2013). *Beyond significance testing: Statistics reform in the behavioral sciences* (2nd ed.). American Psychological Association. https://doi.org/10.1037/14136-000
Lin, M., Lucas, H. C., Jr., & Shmueli, G. (2013). Too big to fail: Large samples and the *p*-value problem. *Information Systems Research, 24*(4), 906–917. https://doi.org/10.1287/isre.2013.0480
Mokkink, L. B., Terwee, C. B., Patrick, D. L., Alonso, J., Stratford, P. W., Knol, D. L., Bouter, L. M., & de Vet, H. C. (2010a). The COSMIN checklist for assessing the methodological quality of studies on measurement properties of health nature measurement instruments: An international Delphi study. *Quality of Life Research, 19*, 539–549. https://doi.org/10.1007/s11136-010-9606-8
Mokkink, L. B., Terwee, C. B., Patrick, D. L., Alonso, J., Stratford, P. W., Knol, D. L., Bouter, L. M., & de Vet, H. C. (2010b). The COSMIN study reached international consensus on taxonomy, terminology, and definitions of measurement properties for health-related patient-reported outcomes. *Journal of Clinical Epidemiology, 63*(7), 737–745. https://doi.org/10.1016/j.jclinepi.2010.02.006
Nakagawa, S., & Cuthill, I. C. (2007). Effect size, confidence interval and statistical significance: A practical guide for biologists. *Biological Reviews, 82*(4), 591–605. https://doi.org/10.1111/j.1469-185X.2007.00027.x
Norman, G. R., Sloan, J. A., & Wyrwich, K. W. (2003). Interpretation of changes in health-related quality of life: The remarkable universality of half a standard deviation. *Medical Care, 41*(5), 582–592. https://doi.org/10.1097/01.MLR.0000062554.74615.4C
Revicki, D. A., Cella, D., Hays, R. D., Sloan, J. A., Lenderking, W. R., & Neil, K. A. (2006). Responsiveness and minimal important differences for patient reported outcomes. *Health and Quality of Life Outcomes, 4*(70), 1–5. https://doi.org/10.1186/1477-7525-4-70
Revicki, D. A., Hays, R. D., Cella, D., & Sloan, J. (2008). Recommended methods for determining responsiveness and minimally important differences for patient-reported outcomes. *Journal of Clinical Epidemiology, 61*(2), 102–109. https://doi.org/10.1016/j.jclinepi.2007.03.012
Terwee, C. B., Roorda, L. D., Knol, D. L., De Boer, M. R., & De Vet, H. C. (2009). Linking measurement error to minimal important change of patient-reported outcomes. *Journal of Clinical Epidemiology, 62*(10), 1062–1067. https://doi.org/10.1016/j.jclinepi.2008.10.011
Tubach, F., Ravaud, P., Martin-Mola, E., Awada, H., Bellamy, N., Bombardier, C., Felson, D. T., Hajjaj-Hassouni, N., Hochberg, M., Logeart, I., Matucci-Cerinic, M., van de Laar, M., van der Heijde, D., & Dougados, M. (2012). Minimum clinically important improvement and patient acceptable symptom state in pain and function in rheumatoid arthritis, ankylosing spondylitis, chronic back pain, hand osteoarthritis, and hip and knee osteoarthritis: Results from a prospective multinational study. *Arthritis Care and Research, 64*, 1699–1786. https://doi.org/10.1002/acr.21747
U.S. Environmental Protection Agency. (2014). *Air Quality Index (AQI) basics*. Retrieved from https://www.airnow.gov/aqi/aqi-basics/
Wasserstein, R. L., Schirm, A. L., & Lazar, N. A. (2019). Moving to a world beyond "$p < 0.05$". *The American Statistician, 73*(sup1), 1–19. https://doi.org/10.1080/00031305.2019.1583913

Chapter 3
Methodological Approaches to Determine MCID and MID: Improve Data-Driven Decisions

Abstract Evaluating treatment efficacy and safety, or clinical endpoints in health technology assessment (HTA), is critical for decision-making in an evidence-based and patient-centered manner. These outcomes should be interpreted in terms of clinical relevance, which is where the minimal clinically important difference (MCID) comes in. Guiding clinical or policy options based on statistical significance alone is bound to fail due to the fact that trivial changes, even with an adequate sample size, can be deemed as significant. Hence, MCID aids in this difficulty by providing the descriptive minimum threshold of change that is regarded by the patients as worthwhile and triggers some form of management change.

Determining the MCID involves two main approaches: anchor-based methods and distribution-based methods. These external references include patient-reported outcomes (PROs) and the clinician's perspective, on which anchor-based methods are reliant. They focus on patients' appreciation of changes deemed as improvements, which can include reduction and betterment of life functions alongside elevation in overall life quality. Unfortunately, this method is affected by personal and cultural differences in the perception of symptoms and change. Taking pain as an example, thresholds differ widely, which makes providing a universal MCID value impossible.

Distribution-based strategies rely on statistical properties, often using a fraction of the standard deviation (SD), like 0.5 SD, as a minimal marker for significant change. Such consistency is advantageous; however, it disregards input from patients, as well as the clinical significance of the findings. Both approaches have been supported by studies, but neither is adequate by itself. As a result, hybrid approaches that integrate qualitative patient input with quantitative data are more comprehensive and thus more effective.

However, there are unresolved issues with MCID, especially in relation to variance. It is bound to a specific context, varies from one medical condition to another, shifts according to patient demographics, changes over time even for the same patient, and is volatile with comorbidities, socioeconomic standing, sex, age, and cultural context. Take, for instance, the meaningful improvement continuum—for a younger reasonable claimant, the yardstick could be entirely different from that set by an older patient with multiple chronic diseases. This conflicts the concept of

I. Fekete, *Transforming Data into Informed Decisions across Clinical and Non-Clinical Domains*, https://doi.org/10.1007/978-3-032-02595-1_3

universal MCID and instead argues the need for tailored or stratified MCID computations.

Notwithstanding its shortcomings, MCID is still critical for harmonizing clinical research with patient-focused care, as well as for contextualizing changes of statistical significance. In this case, however, MCID's usefulness hinges on stride away from uniformity to tailored sensitivity, personalized frameworks, and contextual approaches. This transforms MCID into a flexible instrument able to measure the true clinical impact of health-care interventions in different patient populations and diverse settings.

Keywords Clinical significance · Statistical significance · Effect size · Patient-centered outcomes · PROs · Clinical importance · IQWiG · Distribution- and anchor-based methods

Methodological Approaches to Determine MCID and MID: Improve Data-Driven Decisions

In the world of data-driven decision-making in the clinical field and health technology assessment (HTA), one of the most important tasks is to assess treatment efficacy and harm, sometimes called *effectiveness* and *safety*, respectively (IQWiG, 2023). These two variables are called *clinical endpoints* (IQWiG, 2023). This process is not just a statistical problem; it requires an understanding of both statistics and the subject matter at hand (i.e., the disease), with the latter involving experts from the field (IQWiG, 2023). Hence, we need expert opinions from medical professionals (see, e.g., the Delphi method; Stone Fish & Busby, 2005).

The idea of the Delphi method is that the collective input is more effective than individual expert opinions (Stone Fish & Busby, 2005). It is a structured process designed to gather the views of a group of knowledgeable individuals of the disease (experts in the field) to reach a consensus on a specific topic or aspect of the clinical study (Stone Fish & Busby, 2005).

The treatment could be anything, for example, a new medication for a specific disease. The clinical endpoint (IQWiG, 2023), in statistics language usually referred to as the dependent variable or outcome measure (Field, 2009), could be, for instance, the survival rate of patients, disease progression, days of hospitalization, or any clinical variable of interest, i.e., that we want to explain or predict.

I would like to give some more examples for clinical endpoints from different clinical fields for a better understanding: overall survival (OS), disease-specific survival (DSS), progression-free survival (PFS), recurrence-free survival (RFS), length of hospital stay (LOS), readmission rate, ventilator-free days, time to discharge from hospital, time to disease progression (TTP), tumor response rate (TRR), disease-free survival (DFS), BP reduction, HbA1c levels in diabetes, LDL/HDL, viral load (e.g., HIV RNA copies/mL), C-reactive protein (CRP) levels, liver enzyme normalization (ALT, AST), oxygen saturation or arterial blood gases, pain reduction scores (e.g., measured on VAS), fatigue severity, depression/anxiety scales (e.g., PHQ-9), QoL (e.g., EQ-5D, SF-36), 6-min walk test (6MWT) distance, VO2 max, cognitive function, smoking cessation (abstinence confirmed by CO monitoring), stroke, number of adverse drug reactions (ADRs), serious adverse events (SAEs), dose-limiting toxicities (DLTs), live birth rate, gestational age at delivery, neonatal outcomes, etc. Depending on the clinical indication, specific clinical endpoints should be calculated.

The clinical data can come from various sources, such as clinical trials or even from real-world data (RWE) or already existing registries (for statistical problems and solutions in small registries, see, e.g., Fekete et al., 2020). The data should cover a sufficient period (the observation period depends on the specific clinical indication as per IQWiG guidelines) and include a large enough sample (sample-size calculation is mandatory in clinical trials for early benefit drug assessment) to ensure the quality criteria of the results (IQWiG, 2023). Regrettably, in many studies, no prior sample-size calculation is performed.

Statistical methods are used to determine the efficacy and potential harmfulness of the treatment (IQWiG, 2023). These statistical methods should be *prespecified* as per the German Institute IQWiG (in Germany to give an example) in the so-called statistical analysis plan (SAP) of the clinical study (IQWiG, 2023).

For example, if the difference between the two different groups (i.e., treatment and a control group) is statistically significant, it *does not necessarily entail* that the treatment has efficacy (IQWiG, 2023). In other words, the significant treatment effect is not large enough in terms of the minimal clinically important difference (MCID) (compared to SoC/best practice) for that indication to be perceivable to the majority of patients.

However, there can be a bunch of factors that can confound or contaminate the results. For instance, there could be a so-called "selection bias" where the individuals in the treatment group are, for instance, ethnically different from those in the control group (e.g., only Japanese or only Indian participants were selected), or gender bias, etc. There could also be a placebo effect where the belief that one is receiving treatment leads to an improvement in the outcome.

Furthermore, there could be external factors that affect the outcome but are not included in the analysis, such as self-remission, as in the case of many diseases such as everyday flu or even the serious condition Hepatitis C (HCV). It is estimated that HCV spontaneous clearance ranges from 15% to 36% (Bulteel et al., 2016; Santantonio et al., 2003). Spontaneous resolution/elimination of acute HCV (aHCV) is around 30% according to Santantonio et al. (2003). How can you tell or predict if in one of the groups there are way more patients who experience self-remission[1]? I believe it is impossible, else there would be a test for that. What I can imagine is that a model with a bunch of indication-specific variables is constructed with the clinical endpoint being self-remission.

[1] Spontaneous resolution/elimination.

To address these confounding factors, statistical techniques such as stratification (building strata of patients), regression analysis, or *propensity score matching* (PSM; IQWiG, 2023) are used to counterbalance the two or more different groups at the beginning of the clinical trial. These techniques help to isolate the effect of the treatment from other factors, thereby providing a more accurate estimate of the treatment effect. They also help to counterbalance the two groups. But most importantly, these solutions *do not replace MCIDs*. This is why many times clinical trials have to be repeated in every country in which they strive market access.

A primary problem is the subjective nature of the MCID/MID in my mind. It is a concept that is intrinsically reliant on the individual's viewpoint, hence complicating standardization again. Yet another argument against a universal MCID/MID. However, it may depend on various factors, including gender, age, disease duration, previous medications, current medications, medication tolerance, drug tolerance, comorbidities, ethnicity (Japanese individuals and Indians have different genetics that can affect the efficacy of a drug), as well as patient expectations and compliance.

Furthermore, it can also be contingent on the progression of the disease. Specifically, baseline characteristics encompass all these variables and more, together with the medical condition itself. I advocate for an *individualized nonstandardized MCID* for these and more reasons. A substantial enhancement for one individual [patient-reported outcome (PRO)] may not correspond to the same for another patient. This subjectivity poses an obstacle in establishing a universal and standardized MID, such as the 15% threshold (IQWiG, 2023). The 15% IQWiG-threshold was first formulated in the General Methods version 6.0, published in November 2020.

Moreover, the concept of MCID/MID is not always practical or accurate, especially when dealing with complex or subjective issues such as *pain* (which is highly subjective), emotional health (which fluctuates over time), or QoL (which is a complex umbrella term) in general. There are subjective occasions where the change may be noticeable to the individual, while it may not be objectively measurable in a standardized way, according to IQWiG (IQWiG, 2023), in my opinion.

The notion of MCID/MID has difficulties regarding its adoption and execution (personal communication with medical doctors). Notwithstanding its potential advantages, it is (i) not commonly acknowledged or (ii) employed in medical research or practice, as per their experience. Resistance may arise from medical doctors who favor conventional efficacy measures, qualitative evaluation, or who criticize the subjective aspects of the MCID/MID (personal communication with medical doctors). Furthermore, a lot of medical practitioners are unaware of the existence of the MCID.

Finally, I think it is a challenge that there is also variability over time *within* an individual. The MCID/MID for an individual can change as they adjust to their clinical condition, therapy, or medication over time, for example. What I call temporal variability adds another layer of complexity to MCID and can make it difficult to apply consistently throughout the entire therapy. Just imagine that at the beginning of the therapy, the MCID is the one found in the literature, but with time it diminishes. Accordingly, medication should be adjusted.

Distribution-Based Methods: Standard Error of Measurement, Effect Size, Minimal Detectable Change

MCID denotes the smallest difference or change in a PRO that a patient identifies as meaningful, noticeable, or significant, warranting a change in the treatment plan in the patient's management (e.g., Guyatt et al., 1987; Jaeschke et al., 1989). This threshold serves not only to guide or modify clinical decisions but also to inform the interpretation of clinical trial outcomes toward the authority. Third, it also emphasizes the PROs and puts the patient in the foreground.

The works by Guyatt et al. (1987) and Jaeschke et al. (1989) initially introduced the MCID concept, sparking extensive research and discussion regarding its determination, application, and implications in clinical practice and research. Guyatt et al. (1987) laid the foundation for MCID, emphasizing its utility in discerning clinically meaningful changes from statistical significance.

However, Guyatt et al. (1987) did not explicitly define MCID in their article. Instead, the operational definition of *responsiveness* was employed by them. What is important for us is the responsiveness in measuring the effectiveness of health interventions, emphasizing its significance in identifying changes that are not only *statistically significant* but also *clinically meaningful* to patients.

Jaeschke et al. (1989) continued to elaborate on the MCID concept, proposing methods for its determination, including both so-called "anchor-based" and "distribution-based" approaches. This work highlighted the importance of aligning MCID with patient values and expectations, reinforcing the patient-centered nature of this measure. MCID is conceptualized by Jaeschke et al. (1989) as the most minor variation in health status that patients recognize as beneficial, sufficiently significant to warrant a modification in their management, provided that this change incurs no problematic side effects or disproportionate costs.

Osoba et al. (1998) explored the application of MCID in QoL research, demonstrating the importance of MCID in the context of HRQoL measures. Their work emphasized the subjective nature of MCID. As per their definition, MCID refers to the slightest variation in condition—whether *advantageous* or *detrimental—that a patient can reliably detect* (Osoba et al., 1998). Given the *subjective* nature, I regard this statement again as an argument against a universal MCID/MID.

Similarly, Guyatt et al. (2002) addressed challenges in its application and determination across different settings and outcomes. MCID is defined as the minimal change in a score within the relevant domain that patients deem significant, be it positive or negative (Guyatt et al., 2002). This change is of such a magnitude that it prompts clinicians to contemplate adjustments in patient care strategies. This conceptualization highlights the dual focus on patient perception of change and its clinical implications, bridging subjective patient-experience and objective clinical decision-making.

Schünemann et al. (2005) examined the utilization of MCID in guideline formulation, demonstrating its significance in converting research outcomes into actionable recommendations for patient care. This further highlighted the significance of MCID in connecting research and clinical practice. MCID is defined by them as the minimal change in the outcome measure that is considered meaningful by patients or their representatives (e.g., caregivers). This extent of change is required to achieve for the patient or the practitioner for a possible alteration in the management of the patient's condition. Supporting the nonuniversal nature of MID/MCID, Sloan et al. (2002) highlighted methodological concerns and the variability on MCID estimation.

Similarly to the proposed variability in MCID by Wyrwich et al. (2005), De Vet et al. (2006) conducted a detailed analysis of MCID. Their studies emphasized the fluidity of MCID, caused by contextual variables and patient demographics. A change deemed substantial by patients depends on beginning circumstances or illness severity, the type of intervention, and the duration of the follow-up period, among others. This definition emphasizes the individualized nature of meaningful change, which is consistent with my view if MCID/MID.

Beaton et al. (2001), Beckerman et al. (2001), and Angst et al. (2017) defined MCID differently. In addition, they also considered the measurement error of the instrument employed. This concept of MCID emphasizes the significance of the accuracy of evaluating tools in detecting alterations in individual assessments. The definitions of MCID provided by Beaton et al. (2001), Beckerman et al. (2001), and Angst et al. (2017) all emphasize the patient-centered nature of the metric, influenced by clinical context and measurement tools. This perspective highlights the necessity for adaptable and individual threshold values to maintain the validity and relevance of PROs.

Beckerman et al. (2001) defined the smallest measurable change that exceeds the threshold of random variation or measurement error. This definition highlights that changes are regarded as significant only when they exceed the limits of statistical noise, a definition far from the orthodox definition of MCID. This definition makes it clear that the metric by Beckerman et al. (2001) is not synonymous with MCID. I have included their study because their metric is similar to MID.

The results of an experiment can be affected by random measurement errors, Type I errors (false-positive error), and Type II errors (false-negative error) or even sample size overestimation (Lin et al., 2013). This implies that even differences that are trivial can be accepted as statistically significant (Kalinowski & Fidler, 2010).

Sullivan and Feinn (2012) demonstrate that differences between any means of two groups can become significant if we inflate the sample size large enough, consistent with Lin et al. (2013). I am going to show an experiment from the field of linguistics that did exactly the same malpractice. To avoid this statistical fallacy, we can evaluate the magnitude of differences using MID/MCID, or at least effect size.

MCID measures are usually derived using a standard deviation (SD) criterion and effect size parameters combined, following contemporary guidelines for these methods' triangulation (Crosby et al., 2003; Engel et al., 2018; Revicki et al., 2006,

2008). MID, similar to MCID, is characterized as "the smallest change in score in the construct being measured, which patients perceive as important" (Mokkink et al., 2010).

In this book, I am going to focus on the so-called "distribution-based techniques," which use solely the statistical characteristics of the sample as well as the so-called "anchor-based techniques" (Crosby et al., 2003) because we do not have any anchors in other fields such as the humanities, sociology, linguistics, biology, etc. An "anchor" is basically the patient who can tell about her/his perceptions.

"Distribution-based techniques" have the advantage that specific differences beyond a certain random variation can be identified (Crosby et al., 2003). For example, the one-half standard deviation (SD) criterion of an outcome measure is commonly accepted (e.g., Farivar et al., 2004). This criterion suggests that a score that exceeds half the SD of the outcome is deemed to be important. In other words, we refer to that value as the MCID in a prepost measurement trial (Copay et al., 2007; Farivar et al., 2004; Norman et al., 2003, 2004; Watt et al., 2021). In my opinion, 0.5 SD (Farivar et al., 2004; Norman et al., 2003) is contingent on the patient population and the PRO also (among other factors such as the clinical indication or cultural factors), which might undermine its acceptability.

However, to defend and advocate the 0.5 SD MCID, Norman et al. (2003) analyzed in a systematic review on MIDs 38 studies. Crucially, they found that 32 of these studies reported MIDs approximating half an SD. Norman et al. (2003, 2004) argue for the universality of this threshold (0.5 SD MCID) by presenting psychophysiological evidence. For instance, psychological research (specifically, psychophysics) suggests that across numerous tasks, the average discrimination threshold for humans is 1 in 7, which is about half a standard deviation (Norman et al., 2003, 2004: 582).

This half an SD by (Norman et al., 2003, 2004: 582) deserves some explanation, I should think. The discriminating threshold, sometimes referred to as the barely detectable difference [just noticeable difference (JND)], in perceptual psychology and psychophysics is the smallest difference between two stimuli a person can consistently detect. People sometimes have a heuristic whereby they can detect a difference when the second stimulus differs from the original by around 14–15% threshold, coinciding with the 15% IQWiG threshold in 2023 (my remark). I leave it to the reader to ponder about this alignment between half a SD by Norman et al. (2003, 2004) and 15% (IQWiG, 2023).

This "1 in 7" rule implies that the difference will only be apparent if the comparison weight is at least 114–115 g if a weight is 100 g. This rule holds generally over sensory modalities, including weight, time perception, numerosity, visual brightness, and aural pitch. Crucially, this rule has been adopted by the clinical-psychological field when they administer or administered standardized tests.

In other words, to notice a difference consistently, human perceptual processes thus need a variation of roughly 14% or 0.5 SD between two stimuli. This idea provides a consistent framework to grasp how individuals perceive differences across many contexts.

I wonder silently if there is interindividual variation in this rule (e.g., in the elderly). Therefore, I advocate a personalized MCID/MID throughout the book. I also ask myself if and how other covariates can be incorporated into the calculation of MCID/MID.

While Norman et al. (2004) and Farivar et al. (2004) argue for the 0.5 SD MCID, Eton et al. (2004) propose a 0.3 SD as MCID. Moreover, there are other SD-metrics "on the market," a topic that will be discussed in great detail in the book later. Such uncertainty regarding the SD-criterion speaks, in my view, against its universality.

Vacha-Haase (2001), Norman (2005), Sullivan and Feinn (2012), Lin et al. (2013: 4), and Vasishth and Gelman (2017), among others (including APA, 2010: 33) have suggested reporting effect sizes to exclude significant yet not meaningful differences due to large sample sizes. Namely, large sample sizes "inflate" the *p*-value and give false positive results (so-called Type I error).

But then if we have effect sizes, why do we need MIDs and MCIDs? The reason is that while effect sizes are important, MIDs/MCIDs have to do with the usefulness of an effect. This way, we can check ascertained findings that are in concordance with relevant meaningful outcomes and PROs.

In interpreting the effect-size δ, we can adopt the thresholds set by Romano et al. (2006): $|\delta| < 0.147$ is considered as "negligible," $0.147 \leq |\delta| < 0.33$ as "small," $0.33 \leq |\delta| < 0.474$ as "medium," and $|\delta| \geq 0.474$ as "large" effect-size. In keeping with the notion that small effect sizes denote a nonmeaningful difference and do not meet the MCID/MID criteria, 0.33 may be used as the threshold effect size (Angst et al., 2017; Norman et al., 2003).

Anchor-Based Methods: Patient and Clinician Perspectives

The so-called "anchor-based methodologies" are extensively employed in health sciences to ascertain the MCID. These methods depend on external criteria or "anchors," such as PROs or physician evaluations, to quantify the extent of improvement or harm in a quantifiable construct, such as pain, QoL, or symptom intensity. Alone, the fact that the variability of perception of pain in patients is so high, again, speaks against a universal MCID for pain in this example, in my opinion.

I believe that even a stratification by gender and age would not help given the multiplicity of factors affecting pain tolerance (e.g., cultural factors, individual differences, the disease itself, the severity grade of pain, etc.).

Patients' viewpoints are frequently assessed using PROs, which are *subjective appraisals* of their health condition. These self-reported metrics serve as benchmarks (i.e., MCIDs). A patient may evaluate enhancements in pain or physical function using a Likert scale, for example, which serves as a benchmark for establishing thresholds of clinical significance, that is, the MCID.

Patients present distinct circumstances throughout their assessments. Social, emotional, cultural, and physical factors affect their perception of change. The definition of a significant improvement varies among patients, necessitating the stratification of analyses to represent different patient populations. In my mind, personalized MCIDs, based on subgroup-specific anchors, can improve interpretability and clinical significance.

The significance of the so-called shared decision-making (SDM) in anchor-based thresholds, informed by patient feedback, facilitates the healing process because the clinician gets feedback from the patient (Barry & Edgman-Levitan, 2012).

One example of an anchor is that clinicians provide an objective alternative to PROs. Clinicians utilize instruments like the clinical global impression (CGI) scale to evaluate whether a patient has improved, deteriorated, or maintained stability. This offers external confirmation of PROs, guaranteeing a thorough assessment. Nonetheless, the judgments are susceptible to subjectivity, again, an argument against a universal MCID.

Anchor-based methodologies are essential for determining MCIDs. For example, if patients indicating "slight improvement" equate to a five-point alteration on a pain scale, this serves as a potential threshold for clinical relevance. Incorporating both patient and clinician anchors enhances the reliability of these estimations. "Anchors" mitigate the overinterpretation of insignificant results in clinical trials. However, as already mentioned, anchors may differ as per individual, cultural, ethnic, or socioeconomic circumstances, among others.

Hybrid (Mixed) Methods

Hybrid or mixed methods research integrates qualitative and quantitative methodologies. Mixed methods research refers to both anchor- and distribution-based methods. Anchor-based methods use external criteria (e.g., mainly the patient but also PROs, parents, nurses, or caregiver, etc.). In contrast, distribution-based methods employ statistical features of the data (e.g., the SD-criterion), including effect size or SD, to determine criteria for clinical significance.

Quantitative methods are especially beneficial when there are no anchors available (e.g., Fekete et al., 2018). Hybrid or mixed methods research in MCID entails the systematic mixture of qualitative and quantitative methodologies, which add an extra layer of reliability to the MCID. I would like to emphasize an aspect, namely, that SDM builds trust between clinician and patient.

A qualitative approach, besides a simple anchor-based approach, would encompass, for example, conducting interviews with patients or focus groups to collect subjective insights on significant alterations. By combining clinical insights (qualitative) with evidence-based benchmarks (quantitative), mixed approaches can guarantee that MCID thresholds are both pragmatic and congruent with real-world applications.

This method not only improves the validity and relevance of MCID levels but also ensures they are rooted in the experiences of patients and physicians. As health care increasingly prioritizes PROs, mixed methods research will be essential in reconciling statistical significance with therapeutic relevance.

Criticisms of MCID

While mentioning its value, patient-centered care, MCID has received criticism regarding its usefulness, cross-cultural applicability, subjectivity, methodological rigor, clinical indication-specificity, and context responsiveness. One of the more concerning areas pertains to the lack of attention to individual patient factors such as age, culture, gender, socioeconomic status (SES), disease severity, ethnicity, and comorbidities, in deriving and interpreting MCID values.

Differentiation in age presents the greatest challenge to the use of MCID on a global level. Perceived improvement, symptom tolerance, and the expectation of treatment vary drastically within different age groups. As an example, an elderly patient suffering from chronic arthritis may regard a modest reduction in pain as sufficient for meaningful functional improvement. In contrast, a younger, more active individual would consider a much more substantial change to return to their base level of functioning. Many studies rely on one MCID value for populations, ignoring different age groups. This is especially critical in conditions where there is a significant difference in baseline status of patients' ages, for instance, cardiovascular and neurodegenerative diseases.

Cultural context is perhaps the most neglected aspect in estimating MCID. Culturally relative factors influence how one forms the perception of illness, verbally expresses and experiences symptoms, defines what improvement looks like, and how one interacts with the health-care system. For example, a patient from a collectivist culture may choose to underreport discomfort or downplay treatment failure to fit social expectations, while an individualist would be more inclined to personally assert experiencing treatment dissatisfaction. These variables shape the appraisal of the importance of health change at their individual level. However, most MCID values are based on studies conducted on Western populations and are uncritically adapted to different cultures without validation or modification.

Comorbidities, too, change the applicability of MCID thresholds. Patients dealing with multiple chronic conditions usually have complex and overlapping symptoms, which make it hard to assign any perceived improvement to a single outcome or intervention. For example, a patient with chronic obstructive pulmonary disease (COPD) who is also depressed, and diabetic may assess dyspnea improvement differently than a patient with only pulmonary disease. Hence, the patient's experiences may increase or decrease the subjective threshold of what represents the MCID. In such instances, the rigid application of a fixed MCID value would overlook this multidimensionality of diseases and symptoms.

In addition, gender differences in symptom reporting, health care-seeking behavior, and treatment expectations tend to be ignored when estimating or applying the MCID. Women, for instance, tend to report more pain and fatigue than men and may have different thresholds for defining a clinically meaningful change. Fortunately, in my experience while working in HTA, we stratified patients according to gender and age groups among other variables.

Also, people from lower SES could be more likely to postpone seeking care due to a lack of resources or a distrust in medical institutions, which may enable them to endure higher levels of discomfort before expressing a desire for change. This in turn might change their MCID.

A more subtle yet critical limitation of MCID is its ability to mask interindividual heterogeneity by encapsulating MCID into a single, population-based estimate. While this makes analysis and reporting streamlined, it creates gaps in the framework of individualized health care. Clinical decision-making based on MCIDs runs the risk of overtreating patients who, by virtue of their unique circumstances, may not value marginal improvements, or undertreating patients whose unique circumstances make even minimal changes life-enhancing. This is troubling in areas of mental health, chronic pain, and functional rehabilitation, where patient goals and lived experiences diverge greatly.

In addition, the approach that determines MCID, in particular, seems to overlook the impact of patient-level factors. Distribution-based approaches, which compute measures, for example, 0.5 SD, have no patient input entirely (and are arbitrary thresholds). As a result, such methods can estimate MCID with strong statistical justifications but lack consideration of clinical, desirable, and subjective experiences by the patients.

Unfortunately, as much as the MCID attempts to reconcile clinical research with patient-centered care, it is restrained by a lack of fundamental patient characteristic integration. Patients' age, culture, gender, comorbidities, and SES greatly affect what a meaningful change is from a clinical perspective.

To defend MCID, enhanced, stratified, and personalized approaches need to be adopted, moving away from averages that overshadow the diversity of patients. I advocate a personalized MCID that considers the factors mentioned above.

References

American Psychological Association (APA). (2010). *Publication manual of the American Psychological Association* (6th ed.). APA.

Angst, F., Aeschlimann, A., & Angst, J. (2017). The minimal clinically important difference (MCID) raised the significance of outcome effects above the statistical level, with methodological implications for future studies. *Journal of Clinical Epidemiology, 82*, 128–136. https://doi.org/10.1016/j.jclinepi.2016.11.016

Barry, M. J., & Edgman-Levitan, S. (2012). Shared decision making–The pinnacle of patient-centered care. *New England Journal of Medicine, 366*(9), 780–781. https://doi.org/10.1056/NEJMp1109283

Beaton, D. E., Bombardier, C., Katz, J. N., & Wright, J. G. (2001). A taxonomy for responsiveness. *Journal of Clinical Epidemiology, 54*(12), 1204–1217. https://doi.org/10.1016/s0895-4356(01)00407-3

Beckerman, H., Roebroeck, M. E., Lankhorst, G. J., Becher, J. G., Bezemer, P. D., & Verbeek, A. L. (2001). Smallest real difference, a link between reproducibility and responsiveness. *Quality of Life Research: An International Journal of Quality of Life Aspects of Treatment, Care & Rehabilitation, 10*(7), 571–578. https://doi.org/10.1023/A:1013138911638

Bulteel, N., Partha Sarathy, P., Forrest, E., Stanley, A. J., Innes, H., Mills, P. R., Valerio, H., Gunson, R. N., Aitken, C., Morris, J., Fox, R., & Barclay, S. T. (2016). Factors associated with spontaneous clearance of chronic hepatitis C virus infection. *Journal of Hepatology, 65*(2), 266–272. https://doi.org/10.1016/j.jhep.2016.04.030

Copay, A. G., Subach, B. R., Glassman, S. D., Polly, D. W., & Schuler, T. C. (2007). Understanding the minimum clinically important difference: A review of concepts and methods. *The Spine Journal, 7*(5), 541–546. https://doi.org/10.1016/j.spinee.2007.01.008

Crosby, R. D., Kolotkin, R. L., & Williams, G. R. (2003). Defining clinically meaningful change in health-related quality of life. *Journal of Clinical Epidemiology, 56*, 395–407. https://doi.org/10.1016/S0895-4356(03)00044-1

de Vet, H. C., Terwee, C. B., Ostelo, R. W., Beckerman, H., Knol, D. L., & Bouter, L. M. (2006). Minimal changes in health status questionnaires: Distinction between minimally detectable change and minimally important change. *Health and Quality of Life Outcomes, 4*, 54. https://doi.org/10.1186/1477-7525-4-54

Engel, L., Beaton, D. E., & Touma, Z. (2018). Minimal clinically important difference: A review of outcome measure score interpretation. *Rheumatic Disease Clinics of North America, 44*(2), 177–188. https://doi.org/10.1016/j.rdc.2018.01.011

Eton, D. T., Cella, D., Yost, K. J., Yount, S. E., Peterman, A. H., Neuberg, D. S., Sledge, G. W., & Wood, W. C. (2004). A combination of distribution- and anchor based approaches determined minimally important differences (MIDs) for four endpoints in a breast cancer scale. *Journal of Clinical Epidemiology, 57*, 898–910. https://doi.org/10.1016/j.jclinepi.2004.01.012

Farivar, S. S., Liu, H., & Hays, R. D. (2004). Half standard deviation estimate of the minimally important difference in HRQOL scores? *Expert Review of Pharmacoeconomics & Outcomes Research, 4*(5), 515–523. https://doi.org/10.1586/14737167.4.5.515

Fekete, I., Schulz, P., & Ruigendijk, E. (2018). Exhaustivity in single bare wh-questions: A differential-analysis of exhaustivity. *Glossa: A Journal of General Linguistics, 3*(1), 96. https://doi.org/10.5334/gjgl.549

Fekete, I., Hall, S., Bohn, J., Paradis, A., Schneller, S., Gleißner, E., Coratti, G., Bovis, F., Nascimento, A., Povedano, M., Sormani, M. P., Vázquez, J., Mercuri, E., Kirschner, J., Patel, S. N., & Volmer, T. (2020). Statistical solutions for analytic challenges in registries for HTA-purposes. *Value in Health, 23*(Supplement 2), S692.

Field, A. (2009). *Discovering statistics using SPSS* (3rd ed.). Sage.

Guyatt, G., Walter, S., & Norman, G. (1987). Measuring change over time: Assessing the usefulness of evaluative instruments. *Journal of Chronic Diseases, 40*(2), 171–178. https://doi.org/10.1016/0021-9681(87)90069-5

Guyatt, G. H., Osoba, D., Wu, A. W., Wyrwich, K. W., & Norman, G. R. (2002). Methods to explain the clinical significance of health status measures. *Mayo Clinic Proceedings, 77*(4), 371–383. https://doi.org/10.4065/77.4.371

IQWiG. (2023). *General methods version 7.0*, https://www.iqwig.de/methoden/general-methods_version-7-0.pdf

Jaeschke, R., Singer, J., & Guyatt, G. H. (1989). Measurement of health status: Ascertaining the minimal clinically important difference. *Controlled Clinical Trials, 10*(4), 407–415. https://doi.org/10.1016/0197-2456(89)90005-6

Kalinowski, P., & Fidler, F. (2010). Interpreting significance: The differences between statistical significance, effect size, and practical importance. *Newborn and Infant Nursing Reviews, 10*(1), 50–54. https://doi.org/10.1053/j.nainr.2009.12.007

Lin, M., Lucas, H. C., & Shmueli, G. (2013). Too big to fail: Large samples and the p-value problem. *Information Systems Research, 24*(4), 906–917. https://doi.org/10.1287/isre.2013.0480

Mokkink, L. B., Terwee, C. B., Patrick, D. L., Alonso, J., Stratford, P. W., Knol, D. L., Bouter, L. M., & de Vet, H. C. (2010). The COSMIN checklist for assessing the methodological quality of studies on measurement properties of health nature measurement instruments: An international Delphi study. *Quality of Life Research, 19*, 539–549. https://doi.org/10.1007/s11136-010-9606-8

Norman, G. R. (2005). The relation between the minimally important difference and patient benefit. *COPD: Journal of Chronic Obstructive Pulmonary Disease, 2*(1), 69–73. https://doi.org/10.1081/COPD-200051249

Norman, G. R., Sloan, J. A., & Wyrwich, K. W. (2003). Interpretation of changes in health-related quality of life: The remarkable universality of half a standard deviation. *Medical Care, 41*(5), 582–592. https://doi.org/10.1097/01.MLR.0000062554.74615.4C

Norman, G. R., Sloan, J. A., & Wyrwich, K. W. (2004). The truly remarkable universality of half a standard deviation: Confirmation through another look. *Expert Review of Pharmacoeconomics & Outcomes Research, 4*(5), 581–585. https://doi.org/10.1586/14737167.4.5.581

Osoba, D., Rodrigues, G., Myles, J., Zee, B., & Pater, J. (1998). Interpreting the significance of changes in health-related quality-of-life scores. *Journal of Clinical Oncology: Official Journal of the American Society of Clinical Oncology, 16*(1), 139–144. https://doi.org/10.1200/JCO.1998.16.1.139

Revicki, D. A., Cella, D., Hays, R. D., Sloan, J. A., Lenderking, W. R., & Neil, K. A. (2006). Responsiveness and minimal important differences for patient reported outcomes. *Health and Quality of Life Outcomes, 4*(70), 1–5. :. https://doi.org/10.1186/1477-7525-4-70

Revicki, D. A., Hays, R. D., Cella, D., & Sloan, J. (2008). Recommended methods for determining responsiveness and minimally important differences for patient-reported outcomes. *Journal of Clinical Epidemiology, 61*(2), 102–109. https://doi.org/10.1016/j.jclinepi.2007.03.012

Romano, J., Kromrey, J. D., Coraggio, J., & Skowronek, J. (2006). *Appropriate statistics for ordinal level data: Should we really be using t-test and Cohen's d for evaluating group differences on the NSSE and other surveys?* (pp. 1–3). Paper presented at the annual meeting of the Florida Association of Institutional Research.

Santantonio, T., Sinisi, E., Guastadisegni, A., Casalino, C., Mazzola, M., Gentile, A., Leandro, G., & Pastore, G. (2003). Natural course of acute hepatitis C: A long-term prospective study. *Digestive and Liver Disease: Official Journal of the Italian Society of Gastroenterology and the Italian Association for the Study of the Liver, 35*(2), 104–113. https://doi.org/10.1016/s1590-8658(03)00007-0

Schünemann, H. J., Puhan, M., Goldstein, R., Jaeschke, R., & Guyatt, G. H. (2005). Measurement properties and interpretability of the chronic respiratory disease questionnaire (CRQ). *COPD, 2*(1), 81–89. https://doi.org/10.1081/copd-200050651

Sloan, J. A., Aaronson, N., Cappelleri, J. C., Fairclough, D. L., Varricchio, C., & Clinical Significance Consensus Meeting Group. (2002). Assessing the clinical significance of single items relative to summated scores. *Mayo Clinic Proceedings, 77*(5), 479–487.

Stone Fish, L., & Busby, D. M. (2005). The Delphi Method. In D. H. Sprenkle & F. P. Piercy (Eds.), *Research methods in family therapy* (2nd ed., pp. 238–253). The Guilford Press. https://scholarsarchive.byu.edu/facpub/4584

Sullivan, G. M., & Feinn, R. (2012). Using effect size-or why the P value is not enough. *Journal of Graduate Medical Education, 4*(3), 279–282. https://doi.org/10.4300/JGME-D-12-00156.1

Vacha-Haase, T. (2001). Statistical significance should not be considered one of life's guarantees: Effect sizes are needed. *Educational and Psychological Measurement, 61*, 219–224. https://doi.org/10.1177/00131640121971194

Vasishth, S., & Gelman, A. 2017. The illusion of power: The statistical significance filter leads to overconfident expectations of replicability. In *Proceedings of Cognitive Science Conference*. https://arxiv.org/abs/1702.00556.

Watt, J. A., Veroniki, A. A., Tricco, A. C., et al. (2021). Using a distribution-based approach and systematic review methods to derive minimum clinically important differences. *BMC Medical Research Methodology, 21*, 41. (2021). https://doi.org/10.1186/s12874-021-01228-7

Chapter 4
MCID and MID in Different Clinical Contexts

Abstract One well-known example that illustrates the gap between statistical significance and clinical importance is the antiviral drug called Tamiflu®, for treating influenza. Clinical trials reported a statistically significant reduction in flu symptom duration (around 0.5 days). However, the benefit proved to be clinically irrelevant because the reduction in symptom duration did not warrant the drug's expense, prospective adverse reactions, or the mandatory bulk purchases stemming from policy. This illustrates that the results of statistical analysis, in isolation, are inadequate for assessing actual therapeutic value as well as clinical relevance.

A better description of clinical significance is captured by minimal clinically important difference (MCID), which is defined as the smallest change in a given health condition that a patient considers valuable. MCID denotes not only a change in numbers but also whether that change enhances patients' health, functional status, quality of life, or daily activities. Take, for example, a patient with chronic osteoarthritis of the knee. This patient may not recognize any meaningful improvement even if pain is reduced statistically by 20%. Many chronic pain sufferers expect a $\geq$ 30% reduction before relief is considered worthwhile. Therefore, what is quantifiable through traditional statistical testing might not align with patients' expectations for improvement.

By the disease, treatment, and patient specifics such as age, gender, baseline severity, and expectations, MCID changes. Along with other cultural considerations, due to low dosing strategies aimed at minimizing meaningful change for the elderly, universal thresholds become increasingly difficult to establish. Furthermore, this chapter emphasizes the need for an individualized subjective assessment rather than fixed thresholds.

MCID also aids in patient–physician shared decision-making (SDM), since it permits more effective benchmark-setting. It also helps in ascertaining whether the relevant treatment persists or should be modified to better its outcome, thus offsetting its benefits and strengthening trust between patient and physician.

In the larger contexts of policies concerning health-care systems and drug approval processes, MCID serves as a reference point for assessing cost-effectiveness, reimbursement policies, and intervention prioritization. It enables the precise targeting of health-care spending by identifying interventions that signifi-

I. Fekete, *Transforming Data into Informed Decisions across Clinical and Non-Clinical Domains*, https://doi.org/10.1007/978-3-032-02595-1_4

cantly aid patients, minimizing unwarranted actions, and ensuring that health-care objectives are patient-aligned. Thus, MCID is a pivotal change in moving from a uniform approach using a single benchmark to customized evaluation of health-care interventions.

Keywords Distribution-based methods · Data-driven decisions · Anchor-based methods · Hybrid methods · Standard error · Effect size · Minimal detectable change · MCID

MCID and MID in Different Clinical Contexts

Let me begin with a compelling example where the findings of a study are *statistically significant* but *not clinically relevant*. The antiviral drug *oseltamivir phosphate* (National Center for Biotechnology Information, 2024), with the brand name Tamiflu®, is used for the treatment and prophylaxis of influenza infection viruses A (including pandemic H1N1) and B (National Center for Biotechnology Information, 2024). Tamiflu® was widely promoted for the treatment and prevention of influenza. Several studies (e.g., McClellan & Perry, 2001) claimed that Tamiflu® significantly reduced the duration of flu symptoms by about half a day to a day, based on statistical analyses.

However, when independent researchers reviewed the data, they concluded that while the reduction in symptom duration was statistically significant, its *clinical relevance* was questionable (Jefferson et al., 2014). The minor reduction in symptom duration did not necessarily justify the cost of the medication, the potential side effects (which could include nausea, headache, vomiting, psychiatric and neuropsychiatric effects, or even death, see, e.g., US Food and Drug Administration, 2015, for the side effects), and the public health policy decisions based on its supposed benefits.

Moreover, the review found that Tamiflu® did not significantly reduce the risk of hospitalization or serious influenza complications, which are critical PROs for assessing the clinical relevance of an antiviral drug (IQWiG, 2023; Jefferson et al., 2014).

This discrepancy between statistical significance and clinical relevance initiated a debate on the value of Tamiflu®, which lead to discussions on the importance of assessment of the efficacy and safety of the drug. A treatment can show a statistically significant effect. However, this does not signify that the difference is large enough to have a meaningful and clinically significant impact on (i) patient care, (ii) policymaking, or (iii) public health strategies.

In terms of clinical relevance, operationalized as MCID, cost, policy, therapy, side effects, and risk–benefit ratios are all relevant considerations. Take, for example, the case of Tamiflu®. This case shows how the statistical analysis of medical research requires significant attention with regard to clinical relevance. The greater the clinical significance, the higher the price, to put it simply. But this does not apply to every country. Every country has a different reimbursement system.

MID designates the minimum difference in patient-related outcome (PRO) score that patients find clinically meaningful/noticeable, and without adverse effects and high costs, a change in patient management (Norman et al., 2003). Tamiflu® did not prove to be like one of those.

In contrast to MCID and MID, effect size plays a crucial role in understanding the relationship between clinical importance and PROs. With MCID and MID, effect size helps to determine whether the magnitude of change is not only statistically significant but also clinically meaningful from the patient's perspective (Kazis et al., 1989). My recommendation is to report effect size also as supporting evidence.

MCID Variability Across Diseases and Treatments

Consider the real-life scenario of a patient, a 60-year-old woman with chronic knee osteoarthritis (OA). Despite adhering to a new medication regimen that reduces pain by 20%, the patient finds little solace and reports that the pain has not subsided (despite the *statistically significant* reduction). She still suffers from pain. This highlights again a gap between *statistical significance* and *clinical significance*.

The literature suggests that for chronic pain treatments, patients often seek a 30% (on the 11-point pain scale called PI-NRS; pain intensity numerical rating scale) or greater reduction in pain (Farrar et al., 2001) to achieve *noticeable improvements*. Farrar et al. (2001) used an 11-point numerical rating pain scale. One might ask at this point why there is an 11-point pain scale (PI-NRS). It is because it starts from 0, the highest score (worst) being 10 (we also count 0 as a score). Farrar et al. (2001) revealed also that "higher baseline scores required larger raw changes to represent a clinically important difference." In my view, this makes sense, and, at the same time, supports my view against a universal MCID.

Farrar et al. (2001) used pregabalin.[1] I regard it interesting that they found the same amount of reduction irrespective of the clinical indication (Farrar et al. call them disease type): they studied diabetic neuropathy, chronic low back pain, postherpetic neuralgia, fibromyalgia, and osteoarthritis (OA). Another question I asked myself was, despite the very high number of 2724 participants, why they did not stratify by gender, age, disease type, etc.

For example, Derry et al. (2016) report an even higher value of 41% pain reduction in chronic musculoskeletal pain in adults. Another counterargument of mine against a universal MCID is benzodiazepines (BZDs), which are administered almost in micro-dosages in the elderly for fear of falling and other injuries (Na

[1] Pregabalin is a prescription medicine indicated for alleviating neuropathic pain, epileptic seizures, generalized anxiety disorder (GAD), and fibromyalgia.

et al., 2022). This potential danger applies to tricyclic antidepressants too (Na et al., 2022).

My aim with the example of Derry et al. (2016) was to highlight that baseline characteristics (individual patient characteristics) and clinical indication should be taken into consideration when computing the MCID. These above-mentioned thresholds (not the 15% IQWiG threshold) are supported by PROs indicating that such levels of pain reduction are necessary to significantly improve daily activities and reduce the psychological burden of chronic pain, instead of, say, a 20% pain reduction. Derry et al. (2016) add that "further research may change our estimate of the effect, and very low quality evidence means that we are very uncertain about the accuracy of our estimate," indicating that the topic of MCID definitely needs more research and attention not only within the medical community, but also in nonclinical domains such as the social sciences.

I would like to give an example of another study which clearly shows why MCIDs are needed. As an example, Hara et al. (2022) computed the MCID for PROs of postoperative recovery for gastrointestinal cancer patients in a multicenter study across three Japanese hospitals, which included a sample of 200 patients undergoing elective gastrointestinal cancer surgery. They examined them at baseline and also 4 weeks post-surgery employing health-related quality of life (HRQoL) and walking capacity measures.

Hara et al. (2022) found significant decreases in most health parameters post-surgery, with the SF-36 role-physical (RP) subscale and the 6-min walk test (MWT). This test demonstrated the highest correlation with postoperative recovery. The study concluded that the MCIDs had low predictive accuracy, consistent with my view on the variability and subjectivity of MCIDs.

I contend that this uncertainty is due to interindividual variability, the diversity of the characteristics of diseases, and even the follow-up period, in line with Penning et al. (2023). For example, patient-characteristics/baseline characteristics such as individual age, gender, comorbidities, concomitant medication, severity of the disease, disease duration, therapy resistance, the target organ, and drug tolerance are ignored when proposing MCID/MID (see, e.g., de Vet et al., 2007).

In my mind, even cultural factors can play a role besides previous experiences, mood, and expectations. If a patient has the expectation that the reduction in her/his symptoms will be zero due to the medicine, but the medicine reduces it "only" by 50% (which is, by the way, a success), then we will not reach the MCID because of the high expectations of the patients.

In my opinion, even shame in some cultures can give a false picture of the prevalence and incidence of a disease and the MCID. Some cultures tolerate pain more (Japan and China according to anecdotal evidence), mood can alter the monitoring results in the case of depression, and expectations greatly influence patient satisfaction, which in turn affects the MCID. I claim in the case of progressive diseases that a nonlinear model (such as the one by Wood, 2017) is needed, which considers all the baseline characteristics of the patients as well as the rest of the factors, if they are available.

There are also self-limiting diseases (like influenza) that heal themselves without any medical intervention, as they say. There is also self-remission with many diseases. Some types of depression get better over time as opposed to Alzheimer's disease. These differences in the nonlinear dynamics of disease progression, I think, should also be incorporated into the MCID.

The hypothetical example of this OA patient regarding their experience focuses even more on the value of concentrating the treatment effort on the minimum level needed to achieve a statistically significant and an individual threshold of clinical significance that positively changes the patient's life (Dworkin et al., 2008). So, for example, while a 20% reduction in pain may be 'a step in the right direction,' it is for much of the time insufficient for those with severe chronic pain conditions to enable them to reclaim their lifestyles and daily activities.

Penning et al. (2023) calculated the MCID of the Olerud Molander Ankle Score (OMAS) for patients following surgical treatment for unstable ankle fractures and revealed that MCID values were between 7.0 and 10.5 at 3–6 months and from 5.0 to 7.5 at 6–12 months. One insight from their study is that the *MCID varies depending on the follow-up period* and method used, emphasizing the importance of selecting appropriate MCID values for clinical research.

In sum, the marginal but statistically significant improvement in the hypothetical example did not meet the patient's expectations of pain relief or help to reclaim his or her previous QoL. For patients with chronic pain, achieving a pain reduction of 40% (!) or more is often necessary to significantly improve their QoL (Dworkin et al., 2008). Dworkin et al. (2008) go even as far as claiming that two or more different methods should be used in conjunction to assess the clinical significance of improvement or deterioration for chronic pain clinical endpoints. I definitely agree and can support this standpoint.

MCID is also an instrument for aligning treatment goals with patient expectations (Barrett et al., 2005). It shifts from a "one-size-fits-all" (standardized) approach (such as traditional hypothesis testing or the German 15% standardized threshold defined by the IQWiG, see, IQWiG, 2023) to a more individualized treatment paradigm. A real example of employing MCID is found in the treatment of OA, particularly in the evaluation of knee OA treatments (Lebleu et al., 2024; Tubach et al., 2012): the pain patients experience is high, their mobility is decreased, and QoL is reduced.

Tubach et al. (2012) aimed to determine the threshold for patient-relevant improvement for pain and function in 1362 outpatients with knee OA, i.e., the MCID. Such an MCID would provide a clear threshold for clinical improvements that are noticeable to patients. Tubach and colleagues termed their metric 'MCII' (Minimal Clinically Important Improvement), and not MCID, which does not alter the conclusion of their study, in my opinion. Basically, they speak about the same metric. The authors employed several well-established outcome measures, including the Western Ontario McMaster Universities Osteoarthritis Index (WOMAC), to identify MCII values for pain relief and functional improvement in individuals undergoing treatment for knee OA.

The findings from Tubach et al. (2012) offer valuable insights into the level of improvement that is considered *clinically significant* by patients, which in turn supports clinicians in tailoring treatment plans to meet these expectations and improve patient satisfaction with care. The results by Tubach et al. (2012) indicate that for knee and hip OA, MCII for absolute and relative changes were, respectively, "−40.8% and −32.0% for pain; −39.0% and −32.6% for patient's global assessment; and −26.0% and −21.1% for WOMAC function subscale score."

According to Neumann et al. (2000) *health policy* and *reimbursement decisions* also influence medical interventions and propose *cost-effectiveness analysis* (CEA) as a means of helping policymakers "to determine whether drugs and other interventions offer value for money.". Hence, MCID is instrumentalized in this context to guide policy decisions and reimbursement decisions.

Applying MCID/MID (minimally important difference) enables clinicians to make more better and faster ("faster," an aspect which has not been touched upon so far) decisions regarding treatment options: do they have to continue, adjust, change the drug, or discontinue treatment? This understanding of MCID allows for a more effective allocation of health-care resources and improves the strategic performance of health-care organizations. Hence, employing MCID reduces unnecessary treatments and optimizes PROs, generating business benefits including cost savings and better resource management.

Furthermore, MCID/MID plays a crucial role in influencing health-care policies and insurance coverage decisions. Insurers and policymakers can utilize MCID/MID to determine treatments that are adequately effective for coverage, offering a measurable assessment of treatment efficacy from the patient's viewpoint, which subsequently impacts reimbursement policies and coverage determinations, thereby influencing the financial dynamics of the health-care sector. For example, an insurance company can argue for less reimbursement based on the argument that the PRO did not achieve the MCID for that clinical indication.

In the pharmaceutical sector, MCID/MID is highly essential for clinical trial design and the drug approval process, also known as early benefit assessment of drugs. It facilitates regulatory processes and market access for novel pharmaceuticals and medical devices, allowing companies to demonstrate the efficacy of their products. This is essential for market authorization and competitive advantage.

Health-care providers, pharmaceutical businesses, or medical doctors, who utilize MCID in their communications in every context, can demonstrate their commitment to patient-perceived benefits better than those who do not. In sum, effectively leveraging MCID provides a strategic advantage in the competitive, patient-centric health-care sector.

A clinically relevant difference may not reach statistical significance in a small sample size, for example, in the case of rare diseases such as Spinal Muscular Atrophy (SMA) or Duchenne Muscular Dystrophy (DMD). Fekete et al. (2020) focused on the methodological challenges of, for example, merging clinical practice data from various registries, illustrated by the case of 5q-associated SMA. The challenges encompass missing data, inconsistent data quality, imbalance in treatment groups, confounding factors, nonparametric distributions, patient heterogeneity, and small patient registries. These challenges arise because patient characteristics cannot be controlled in RWE.

The misinterpretation, miscalculation, or misalignment of *statistical* and *clinical significance* can lead to flawed decision-making in clinical studies. For instance, a treatment that shows a statistically significant improvement in laboratory results may not lead to a meaningful improvement in the patient's QoL, rendering it *clinically irrelevant*.

I must admit MCID does not apply to all medical conditions. In many cases (e.g., antibiotics), patients do not even feel anything, they might not even have any symptoms at all. Still, they have to take antibiotics. In such cases, I believe it makes no sense, or it is even impossible to calculate MCID.

Some clinical trials do not report any MCIDs despite the fact that they have large databases (e.g., Koeks et al., 2017). For example, Koeks et al. (2017) studied DMD and aimed to evaluate the impact of corticosteroids on key clinical outcomes in a large global cohort of genetically confirmed DMD patients.

Their analysis encompassed data from a very large cohort of 5345 genetically confirmed DMD patients across 31 countries. They used stratification based on age and corticosteroid usage. The findings by Koeks et al. (2017) highlight in DMD patients that there are beneficial effects of corticosteroid on ambulation, scoliosis surgery, ventilation, and cardiomyopathy. However, the question arises regarding why they did not use any MCID. It would have bolstered their findings.

Duong et al. (2021) also focused on DMD. With the possible treatments aimed at muscle weakness and wasting, it is important to define clinically relevant change in functional outcomes precisely. Their retrospective analysis derived from the multicenter natural history study sought to determine among other things the so-called minimal detectable change (MDC) and the MCID of 12-month changes in so-called standardized timed function tests (TFT), which have been designed for monitoring the progression of DMD.

Duong et al. (2021) identified significant MDC and MCID values for three commonly used TFTs: (i) supine to stand, (ii) 10 m walk/run, and (iii) four stair climb. However, I must mention that the effect sizes that their study yielded were close to or greater than 0.2. Their findings underscore the *clinical relevance* of these TFTs in detecting meaningful changes in DMD progression, as per their report (Duong et al., 2021).

The Role of Informed Decisions for Patient Care

MCID is important to informed decisions for patient care, since it guides patients and physicians in assessing *together* the efficacy (or harm) of a therapeutic effect, earlier referred to as *shared decision-making* SDM. Both of them can even *track the improvement using MCID*. I would like to add that, in my view, beside these, SDM builds trust in the doctor and also in the therapy.

A patient-centered metric is obtained from clinical research (MCID), usually with the use of distribution-based or anchor-based techniques (for a systematic review on the anchor-based approaches, see Zhang & Huang, 2023). Anchor-based approaches assess how the outcome measure compares to an external threshold (or metaphorically speaking, an "anchor"). It can be a clinical judgment by an MD, a nurse, or a care-giver, or the patient, etc.

The idea of SDM approaches is that patients and physicians choose treatment alternatives together. As already mentioned, it also establishes trust between patient and doctor. For instance, the doctor can openly convey that the benefit might not live up to the patient's expectations if an OA medicine produces a 10% improvement in function but the MCID for functional improvement is 15% (let us take the German threshold for the time being, IQWiG, 2023, p. 173).

All in all, these conversations enable patients to make decisions that suit their preferences. In addition, SDM also facilitates clinicians' confidence in the efficacy of the therapy by strengthening the communication that patients have with their medical practitioners.

MCID is very helpful when it comes to therapy customization and modification (e.g., drug adjustment). Individual thresholds for seeing clinically significant change, comorbidities, and baseline conditions vary through all patients. As already mentioned, stratifying MCIDs according to patient subgroups, such as age, condition severity, or baseline function can improve personalized MCID estimation. My suggestion is that MCIDs can be utilized also to track the progression of a disease. To illustrate, if a patient has exceeded the MCID, then it can be conceived of as feedback for the patient. As a result, the patient will be happy, which might affect their QoL. Let us think about the 6MWT (6-min walk test in meters).

To illustrate, stratified and individualized MCIDs are necessary because, for example, elderly patients with chronic obstructive pulmonary disease (COPD) may have different MCIDs than younger patients. The same applies to almost all diseases. Hence, patient characteristics should be considered when computing the MCID for the clinical indication.

Patients must be fully informed about the possible advantages, risks, and available options of a treatment in order to comply with the principle of informed consent. Patients are guaranteed to comprehend the practical effects of an intervention when MCID/MID is incorporated into this procedure. Before agreeing to treatment, patients have the right to know, for example, if a new drug only provides a slight improvement that falls short of the MCID criteria.

Admittedly, there are certain difficulties with MCIDs. First of all, MCIDs are indication-specific; they depend on the patient demographic and on baseline characteristics of the patient. Finally, another problem is that not all clinical trials record MCIDs, which makes it difficult to compare studies.

References

Barrett, B., Brown, R., Mundt, M., Safdar, N., Dye, L., Maberry, R., & Alt, J. (2005). The Wisconsin upper respiratory symptom survey is responsive, reliable, and valid. *Journal of Clinical Epidemiology, 58*(6), 609–617. https://doi.org/10.1016/j.jclinepi.2004.11.019

de Vet, H. C., Ostelo, R. W., Terwee, C. B., van der Roer, N., Knol, D. L., Beckerman, H., Boers, M., & Bouter, L. M. (2007). Minimally important change determined by a visual method integrating an anchor-based and a distribution-based approach. *Quality of Life Research: an International Journal of Quality of Life Aspects of Treatment, Care and Rehabilitation, 16*(1), 131–142. https://doi.org/10.1007/s11136-006-9109-9

Derry, S., Conaghan, P., Da Silva, J. A., Wiffen, P. J., & Moore, R. A. (2016). Topical NSAIDs for chronic musculoskeletal pain in adults. *The Cochrane Database of Systematic Reviews, 4*(4), CD007400. https://doi.org/10.1002/14651858.CD007400.pub3

Duong, T., Canbek, J., Birkmeier, M., Nelson, L., Siener, C., Fernandez-Fernandez, A., Henricson, E., McDonald, C. M., Gordish-Dressman, H., & Investigators, C. I. N. R. G.-D. N. H. S. (2021). The minimal clinical important difference (MCID) in annual rate of change of timed function tests in boys with DMD. *Journal of Neuromuscular Diseases, 8*(6), 939–948. https://doi.org/10.3233/JND-210646

Dworkin, R. H., Turk, D. C., Wyrwich, K. W., Beaton, D., Cleeland, C. S., Farrar, J. T., Haythornthwaite, J. A., Jensen, M. P., Kerns, R. D., Ader, D. N., Brandenburg, N., Burke, L. B., Cella, D., Chandler, J., Cowan, P., Dimitrova, R., Dionne, R., Hertz, S., Jadad, A. R., Katz, N. P., et al. (2008). Interpreting the clinical importance of treatment outcomes in chronic pain clinical trials: IMMPACT recommendations. *The Journal of Pain, 9*(2), 105–121. https://doi.org/10.1016/j.jpain.2007.09.005

Farrar, J. T., Young, J. P., Jr., LaMoreaux, L., Werth, J. L., & Poole, M. R. (2001). Clinical importance of changes in chronic pain intensity measured on an 11-point numerical pain rating scale. *Pain, 94*(2), 149–158. https://doi.org/10.1016/S0304-3959(01)00349-9

Fekete, I., Hall, S., Bohn, J., Paradis, A., Schneller, S., Gleißner, E., Coratti, G., Bovis, F., Nascimento, A., Povedano, M., Sormani, M. P., Vázquez, J., Mercuri, E., Kirschner, J., Patel, S. N., & Volmer, T. (2020). Statistical solutions for analytic challenges in registries for HTA-purposes. *Value in Health, 23*(Supplement 2), S692. https://doi.org/10.1016/j.jval.2020.08.1750

Hara, T., Kogure, E., Iijima, S., et al. (2022). Minimal clinically important difference in post-operative recovery among patients with gastrointestinal cancer. *Support Care Cancer, 30*, 2197–2205. https://doi.org/10.1007/s00520-021-06632-9

IQWiG. (2023). *General methods version 7.0*, https://www.iqwig.de/methoden/general-methods_version-7-0.pdf

Jefferson, T., Jones, M. A., Doshi, P., Del Mar, C. B., Hama, R., Thompson, M. J., Spencer, E. A., Onakpoya, I., Mahtani, K. R., Nunan, D., Howick, J., & Heneghan, C. J. (2014). Neuraminidase inhibitors for preventing and treating influenza in adults and children. *The Cochrane Database of Systematic Reviews, 2014*(4), CD008965. https://doi.org/10.1002/14651858.CD008965.pub4

Kazis, L. E., Anderson, J. J., & Meenan, R. F. (1989). Effect sizes for interpreting changes in health status. *Medical Care, 27*(3 Suppl), S178–S189. https://doi.org/10.1097/00005650-198903001-00015

Koeks, Z., Bladen, C. L., Salgado, D., van Zwet, E., Pogoryelova, O., McMacken, G., Monges, S., Foncuberta, M. E., Kekou, K., Kosma, K., Dawkins, H., Lamont, L., Bellgard, M. I., Roy, A. J., Chamova, T., Guergueltcheva, V., Chan, S., Korngut, L., Campbell, C., Dai, Y., et al. (2017). Clinical outcomes in Duchenne muscular dystrophy: A study of 5345 patients from the TREAT-NMD DMD global database. *Journal of Neuromuscular Diseases, 4*(4), 293–306. https://doi.org/10.3233/JND-170280

Lebleu, J., Daniels, K., Pauwels, A., Dekimpe, L., Mapinduzi, J., Poilvache, H., & Bonnechère, B. (2024). Incorporating wearable Technology for Enhanced Rehabilitation Monitoring after hip and knee replacement. *Sensors (Basel, Switzerland), 24*(4), 1163. https://doi.org/10.3390/s24041163

McClellan, K., & Perry, C. M. (2001). Oseltamivir: A review of its use in influenza. *Drugs, 61*(2), 263–283. https://doi.org/10.2165/00003495-200161020-00011

Na, I., Seo, J., Park, E., & Lee, J. (2022). Risk of falls associated with long-acting benzodiazepines or tricyclic antidepressants use in community-dwelling older adults: a nationwide population-based case–crossover study. *International Journal of Environmental Research and Public Health, 19*(14), 8564. https://doi.org/10.3390/ijerph19148564

National Center for Biotechnology Information. (2024). *PubChem Compound Summary for CID 65028, Oseltamivir*. Retrieved March 23, 2024 from https://pubchem.ncbi.nlm.nih.gov/compound/Oseltamivir

Neumann, P. J., Sandberg, E. A., Bell, C. M., Stone, P. W., & Chapman, R. H. (2000). Are pharmaceuticals cost-effective? A review of the evidence. *Health Affairs, 19*(2), 92–109. https://doi.org/10.1377/hlthaff.19.2.92

Norman, G. R., Sloan, J. A., & Wyrwich, K. W. (2003). Interpretation of changes in health-related quality of life: The remarkable universality of half a standard deviation. *Medical Care, 41*(5), 582–592. https://doi.org/10.1097/01.MLR.0000062554.74615.4C

Penning, D., Kleipool, S., van Dieren, S., et al. (2023). The minimal clinically important difference (MCID) of the Olerud Molander Ankle Score (OMAS) in patients with unstable ankle fracture. *Archives of Orthopaedic and Trauma Surgery, 143*, 3103–3110. https://doi.org/10.1007/s00402-022-04533-y

Tubach, F., Ravaud, P., Martin-Mola, E., Awada, H., Bellamy, N., Bombardier, C., Felson, D. T., Hajjaj-Hassouni, N., Hochberg, M., Logeart, I., Matucci-Cerinic, M., van de Laar, M., van der Heijde, D., & Dougados, M. (2012). Minimum clinically important improvement and patient acceptable symptom state in pain and function in rheumatoid arthritis, ankylosing spondylitis, chronic back pain, hand osteoarthritis, and hip and knee osteoarthritis: Results from a prospective multinational study. *Arthritis Care and Research, 64*, 1699–1786. https://doi.org/10.1002/acr.217471625-1786

U.S. Food and Drug Administration. (2015). Tamiflu pediatric adverse events: Questions and answers.

Wood, S. N. (2017). *Generalized additive models: An introduction with R* (2nd ed.). Chapman and Hall/CRC. https://doi.org/10.1201/9781315370279

Zhang, Y., Xi, X., & Huang, Y. (2023). The anchor design of anchor-based method to determine the minimal clinically important difference: A systematic review. *Health and Quality of Life Outcomes, 21*(1), 74. https://doi.org/10.1186/s12955-023-02157-3

Chapter 5
Advanced Statistical Methods in MCID/MID Research

Abstract The determination of the minimal clinically important difference (MCID) and minimal important difference (MID) is essential for evaluating the significance of changes in patient-reported outcomes (PROs) in clinical research and practice. These thresholds help distinguish between statistically significant results and those that are, in reality, meaningful and noticeable/perceivable to patients. This chapter introduces the three main methodological approaches to determining MCID: distribution-based methods, anchor-based methods, and hybrid methods that combine both.

Distribution-based methods rely solely on statistical properties of the data. Key techniques include the standard error of measurement (SEM), effect size (ES), and minimal detectable change (MDC). SEM quantifies how much variation in scores is due to measurement error and helps identify real changes in individual patient scores. MDC, derived from SEM, establishes a threshold beyond which a change is unlikely to be due to random variability. Effect size, often measured with standardized differences such as Cohen's *d*, provides the magnitude of change, which is categorized as small, medium, or large. These methods, albeit robust, lack clinical context and other demographic and clinical variables. Hence, they must be interpreted with caution.

Anchor-based methods, in contrast, rely on external criteria—such as patient, caregiver, or clinician ratings—to determine what constitutes a meaningful and noticeable change. For example, a patient might rate their current health compared to an earlier point using a global rating of change scale. This subjective input is crucial in capturing what patients perceive as improvements or deteriorations in their condition. Clinician evaluations also serve as valid anchors, especially when determining the relevance of changes across various domains of functioning.

Hybrid methods combine the strengths of both distribution- and anchor-based approaches. This allows researchers to triangulate findings and produce more comprehensive and reliable MCID estimates. By integrating statistical evidence with patient and clinician perspectives, hybrid methods increase the validity and applicability of the results.

Advanced statistical models such as multivariate regression, linear mixed effects models (LMEM), and generalized additive mixed models (GAMM) enhance the

I. Fekete, *Transforming Data into Informed Decisions across Clinical and Non-Clinical Domains*, https://doi.org/10.1007/978-3-032-02595-1_5

estimation of MCID, especially in longitudinal studies. These models account for repeated measures and interindividual variability, making them ideal for tracking patient progress over time. Techniques like structural equation modeling (SEM) also allow researchers to analyze complex relationships between latent constructs and observed outcomes.

Meta-analytic techniques further refine MCID estimates by synthesizing data across studies. Pooling standard deviations and PROs from various studies ensures broader generalizability. However, heterogeneity in patient populations, measurement instruments, and methods can pose challenges.

Ultimately, determining MCID is not a one-size-fits-all process. It varies by patient characteristics, conditions, and instruments used. A thoughtful combination of methodologies ensures that the MCID reflects both statistical rigor and clinical relevance, fostering patient-centered care and more effective clinical decision-making.

Keywords Distribution-based methods for MCID · Standard error of the mean · Effect size · Minimal detectable change · Anchor-based methods for MCID · Change difference · Standard error of measurement · Regression analysis · Patient perspectives in anchor-based methods · Clinician assessments in anchor-based methods · Synthesizing MCID data

Enhanced Statistical Methods in MCID/MID Research

The determination of MCID/MID is crucial in clinical research and practice, as these metrics help in interpreting the importance and relevance of changes in PROs. Understanding these concepts can lead to improved data-driven decisions in health care in that they improve patient care and resource allocation and even cost determination. This chapter introduces advanced methodological techniques to determine MCID and MID, focusing on so-called distribution-based methods, anchor-based methods, and hybrid (mixed) methods.

Distribution-based methods are statistical approaches that utilize only the distribution of scores from a sample to determine the MCID or MID. These methods often rely on metrics such as the standard error of measurement (SEM), effect size (ES), and minimal detectable change (MDC).

The standard error of the mean, also abbreviated as SEM (and not to be confused with structural equation modeling), is a critical statistical measure that quantifies the amount of error inherent in an individual's score due to measurement variability. It is calculated using the standard deviation of the scores and the reliability of the measurement instrument. The SEM provides a range within which an individual's true score is likely to fall, allowing researchers to determine whether the observed changes in scores exceed this error range. This error range indicates a clinically meaningful change (Zhang et al., 2012).

Often, the application of SEM to ascertain MCID combines both clinical and statistical methods. Gupta et al. (2023) highlighted both SEM and distribution-based approaches in calculating the MCID of NSAA (North Star Ambulatory Assessment) in boys 7–10 years old with Duchenne muscular dystrophy (DMD).

Gupta et al. (2023) calculated an MCID as 2.3–2.9 points for NSAA using 1/3 standard deviation contrary to the most widespread 0.5 SD. Gupta et al. (2023) noted that SEM appears to provide a slightly higher range of 2.9–3.5 points, which hints that *SEM calculations are very conservative*.

Mouelhi et al. (2020) combined the so-called anchor- and distributional-based methods to calculate MCID. They argue that no one method is perfect. However, the combination of statistical approach and clinical perception is fair to determine the MCIDs.

In addition, Sobreira et al. (2020) present another case where they calculated the MCIDs associated with pain, pulmonary function, and the general activity level of patients with spinal cord injury using SEM. They provided threshold values and determined SEM as part of calculating MDC for the patient's treatment and experience (Sobreira et al., 2020). Their study illustrates the importance of SEM regarding the appreciation of the patient's experience relative to actual change in clinical practice.

Another example for SEM is Zhu et al. (2023) who focused on patients with heart failure (HF), wherein they set MCIDs, using distribution-based methods to derive SEM. They pointed out that comprehension above the MCID established by SEM was clinically useful, reinforcing the construct claim of SEM as an evaluation tool in assessing treatment efficacy (Zhu et al., 2023).

Gupta et al. (2023) used distributed structural equation modeling to estimate the MCID for the North Star Ambulatory Assessment in 7–10 year-old boys with DMD. With regard to the SEM estimate, the authors of the study provided the NSAA with an MCID range of 2.3–2.9 points using 1/3 SD. Gupta et al. (2023) indicate that SEM provided an upper limit range of 2.9–3.5 for the MCID estimate. The analyses suggest that the models offer conservative estimates which are useful for encouraging clinical action (Gupta et al., 2023).

Importantly, Mouelhi et al. (2020) explored a variety of methods to determine the MCID and underscored the importance of *combining both anchoring and distribution techniques, as in SEM*, to strengthen the results.

Not long ago, Sobreira et al. (2020) used structural equation modeling (also abbreviated as SEM) to utilize pain, lung functions, and overall functioning in people with spinal cord injury to estimate the MCID, and they pointed out a novel example. The authors highlighted the role of SEM in determining the least detectable change (MDC) and provided values that were meaningful concerning the patients' perspectives and experiences regarding the treatment outcomes (Sobreira et al., 2020).

I would like to stress at this point already that both Bayesian statistics and non-Bayesian statistics (after p-value correction) combat Type error (false positives) but cannot tackle the problem that is raised by this book, namely the problem of the practical implication of a genuinely significant difference.

Another example of the use of SEM is in the study by Zhu et al. (2023) that focused on patients with heart failure. Zhu et al. (2023) reinforced the application of SEM in treatment effect evaluation by stating that modifications above the MCID estimated with SEM were deemed clinically important (Zhu et al., 2023).

Consistently, Wu et al. (2020) also illustrate the relevance of SEM in estimating MCID for chronic diseases like esophageal cancer, where they studied SEM. Wu et al. (2020) explained that SEM's lack of dependence on identical circumstances for estimation within multiple empirics helps with differentiating clinical change from measurement error (Wu et al., 2020).

Effect size is another distribution-based method that quantifies the magnitude of change in a treatment group relative to a control group. Cohen's *d* is frequently used as a measure which is the difference between two means divided by the pooled SD. A larger effect size indicates a more substantial difference, which can inform clinicians about the practical significance of treatment effects. For instance, an effect size of 0.2 is considered small, 0.5 medium, and 0.8 large, providing a framework for interpreting the clinical relevance of findings.

Minimal detectable change (MDC) refers to the smallest change in a measurement that reflects a true change rather than measurement error. It is calculated based on the SEM and provides a cutoff that must be exceeded so that we can say that a change or difference is clinically significant. This approach is very useful in rehabilitation because patient tracking and patient progress are very important. Hence, clinicians can make informed decisions about treatment efficacy and PROs by defining an MDC.

In contrast, anchor-based approaches use external criteria or "anchors" as MCID or MID (e.g., Mouelhi et al., 2020). These anchors can be PROs, clinician assessments, MDs, caregivers, or other relevant measures that provide context or feedback for interpreting changes in health status.

I outline the multiplicity of distribution- and anchor-based methods for the calculation of MCID/MID (Fig. 5.1) relying on the review by Mouelhi et al. (2020). The best solution is to combine both distribution- and anchor-based approaches.

Various Metrics for MCID/MID

Distribution-centric methodologies encompass (Mouelhi et al., 2020):

1. Standard deviation (SD): MCID is computed as a multiple of the SD of the mean change, with 0.5 SD being the most widely used benchmark (used in the majority of studies, 23 studies, as revealed by Mouelhi et al. (2020).
2. Standard error of measurement (SEM): MCID is defined as multiplying the SEM of the HRQoL change score by 1.96 (Mouelhi et al., 2020). A change exceeding the baseline SEM is regarded as a genuine change. Actually, "MDC is calculated using the SEM (Mouelhi et al., 2020, p. 3)." SEM is not a definition of MCID.

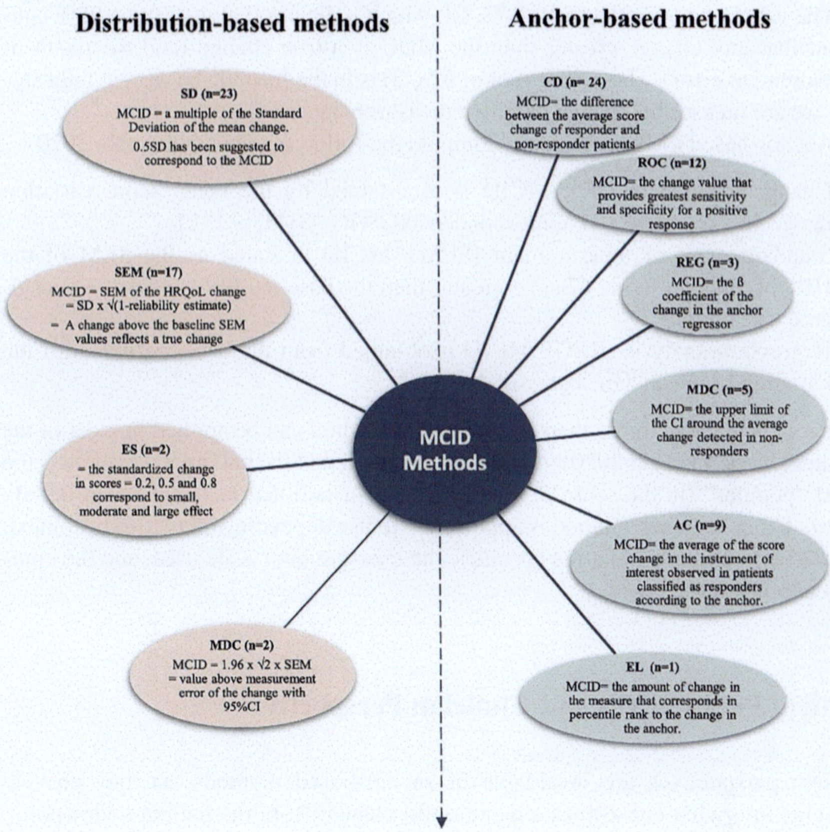

Note: more than one statistical method was used in most studies.

MCID: Minimal Clinically Important Difference, AC: Average Change, MDC: Minimal Detectable Change, CD: Change Difference, ROC: Receiver Operating Curve, REG: Regression analysis, EL: Equipercentile Linking, SD: Standard deviation, SEM: Standard error of measurement, ES: Effect size

Fig. 5.1 A review of anchor-based and distribution-based methods by Mouelhi et al. (2020). It can clearly be seen that half an SD criterion is the most widespread threshold among the distribution-based criteria.

3. Effect size (ES): This denotes the standardized alteration in scores, with values of 0.2, 0.5, and 0.8 indicating modest, moderate, and significant effects, respectively. Crucially, ES is not a definition for MCID.
4. Minimal detectable change (MDC): The Minimal Clinically Important Difference (MCID) is determined by multiplying the standard error of measurement (SEM) by 1.96 to accommodate measurement error, with a 95% confidence interval (CI) as per Mouelhi et al. (2020). MDC=SEM*1.96*$\sqrt{2}$ (Mouelhi et al., 2020). As already mentioned, "MDC is calculated using the SEM (Mouelhi et al., 2020, p. 3)." Crucially, MDC is not a formula for MCID.

The value of 1.96 reflects the 95% CI, which indicates that we can be 95% confident that any change greater than the MDC is a true change (and the result of measurement error). The square root of 2 ($\sqrt{2}$) is in the formula because it indicates that we are measuring change with two measurements.

Anchor-based methodologies encompass the following (Mouelhi et al., 2020):

1. Change difference (CD): MCID is ascertained by the mean score variation between responders and nonresponders (IQWiG, 2023, p. 172).
2. Standard error of measurement (SEM): MCID is stated as the SEM of the HRQoL change score. Change greater than the baseline SEM would be considered true change.
3. Regression analysis (REG): MCID is obtained from the beta coefficient of the variation in the anchor regressor.

As shown above, more than one statistical method can be applied in most of the studies (Mouelhi et al., 2020). Also, it can be noted that the half an SD method is the most "popular" (in the sense of widespread) distributional metric (Mouelhi et al., 2020). Crucially, each method is appropriate to use depending on the study context, the disease and the population specifics, the measurement scale used, and the clinical endpoints of interest.

Patient Perspectives and Clinician Perspectives

Patient perspectives are invaluable in anchor-based methods, as they provide insights into what constitutes a meaningful change from the patient's viewpoint. For instance, patients may be asked to rate their perceived change in health status on a scale (e.g., "much worse" to "much better"). This subjective assessment can thus be combined with the objective measures, and clinically important difference thresholds are set. Incorporation of patient perspectives is known to make MCID values that are much more practical and enhances the patient centeredness of care.

Similarly, anchor methods rely heavily on clinician assessments. With his multidisciplinary knowledge and experience, a clinician is in a position to determine the degree of patient progress. Such clinician-rated changes can be compared with objective measures of outcome to determine what constitutes a clinically important difference. The combination of the patient and clinician perspectives offers an expanded view of the changes in health outcomes that matter most (Byun et al., 2013).

Hybrid (Mixed) Methods

Unsurprisingly, hybrid methods combine both distribution-based and anchor-based approaches to provide a more robust determination of MCID and MID. This mixed-methods approach (hybrid) leverages the strengths of both methodologies, enhancing the validity and reliability of the findings.

By integrating distribution-based and anchor-based methods, researchers can triangulate data to establish a more comprehensive understanding of clinically important differences. For example, a study might use SEM to calculate the MDC while simultaneously gathering patient-reported outcomes to identify perceived changes.

Hybrid methods integrate both distribution-based and anchor-based methodologies to characterize and strengthen the determination of MCID and MID. The increased accuracy of the findings is achieved through the mixed-methods approach which optimally combines both methods.

While defining clinically important differences, it is important for researchers to employ both the traditional and contemporary data methodologies to better understand the problem being researched. For example, the research may apply invalidity SEM to determine the minimum detectable change (MDC) while concurrently utilizing patient-reported outcomes for supposed changes.

Through the combination of statistical procedures and patient outcomes, a more precise understanding of the effects of the intervention on the patient's health is possible. In addition, this strategy increases the strength of the investigation and contributes to evidence-based practices in health care (Aarons et al., 2017).

I will now proceed to the one with a very high face validity as the authors themselves are willing to claim; this is what they refer to as global rating of change (GRC) which attempts to capture the minimal important change or difference in the patients (Kamper et al., 2009). Patients have to compare their present health status with a specific earlier point in time, for example, then mark the difference on an appropriate scale (for a review of GRC, see Kamper et al., 2009).

The magnitude of this difference is subsequently measured by means of either a numerical or visual scale. In my experience, participants prefer the visual analogue scale (VAS) over the rest of the scales, for example, the Likert scale. The GRC is fast and easy to deploy and gives information on a patient's health in an effective manner (Kamper et al., 2009). The GRC can be used to monitor, for example, the motor-functional status of a patient or a patient with depression. To the best of my knowledge, MCIDs are computed mainly in HTA and at the beginning and at the end of a study.

Various age groups, genders, and severity of illness can influence the MCID. MANOVA or multivariate regression can identify subgroup MCIDs by evaluating predictors against the outcomes, determining specific MCIDs for those groups.

PCA and factor analysis differ by allowing the researcher to narrow their focus on major variables, thereby reducing data complexity for specific change indicators. This is ideal for variables measured through multi-item scales such as QoL.

At times, physical functioning and mental well-being are looked at in a single entity in clinical trials. In this case, SEM or canonical correlation analysis can examine the correlation of different outcomes and provide a single MCID that is applicable to all outcomes.

The GEE model is one of the methods that can provide better and more accurate MCID estimates by taking into consideration the change in functional status over time, rather than a single instance. This method, like the Zhang Method, allows for the use of multivariate mixed effects models, estimating changes over repeated measure data while accounting for correlation between measures.

As MCID is ideally not constant and differs for individuals, distribution-based techniques for estimating it (e.g., half a standard deviation) should be used. Combined with multivariate regression, this method takes individual variability into account.

Accounting for confounding factors can help improve accuracy because they capture the variation of different patient groups. Multivariate techniques provide a more advanced way of dealing with confounding factors.

MCID is very important in measuring the effectiveness of a clinical trial intervention with the aid of statistical models such as linear mixed effects models (LMEM; Pinheiro et al., 2018) or generalized additive mixed models (GAMM; Wood, 2017).

With specific regard to LMEM (e.g., used by Peeters et al., 2023) and GAMM, the advantages from these models come in the form of dealing with longitudinal data with repeated measures taken from the same subjects over time, which is common in clinical trials. It is also typical in chronic disease conditions like Alzheimer's disease where an individual assessment of treatment effectiveness and the correlation of repeated measures has to be done (Hanff et al., 2024; Schober & Vetter, 2021).

Let us consider a clinical trial that assesses a novel cognitive intervention in people diagnosed with advanced stage Alzheimer's disease, where investigators might apply LMEM to evaluate change in cognitive function assessments at different intervals. The model would allow one to measure fixed variables (such as treatment and control groups), as well as random effects (e.g., individual differences and other confounding factors). This allows for a greater insight of how the intervention in question modifies cognition functions over time (Hanff et al., 2024; Schober & Vetter, 2021).

In accounting for the MCID change for cognitive function scores, it is also possible in a trial to determine whether these changes are statistically and clinically significant for the patients. For instance, in the event LMEM determines that there is a mean increase of 2.5 in the scores of a cognitive assessment tool, but this is lower than the MCID threshold of 3, then it is said that the result, even though statistically significant, is clinically insignificant (Bloom et al., 2021).

Further, GAMM (Wood, 2017) can also be used to model nonlinear relationships in the data, which can be particularly beneficial for capturing the effect of time or treatment that may deviate from a linear progression. Such flexibility enables investigators to delineate intricate details of patient reactions that are usually overlooked by ordinary linear models (Hanff et al., 2024; Schober & Vetter, 2021). For instance, if cognitive function enhancement increases at a nonlinear rate in which patients display breakthroughs followed by a period of dynamic equilibrium, it can be simulated using nonlinear GAMM (Wood, 2017), which helps to explain when and how much clinically worthwhile enhancement happens (Peeters et al., 2023).

The strategy of using MCID in conjunction with LMEM and GAMM in clinical trials not only builds the treatment effect relevance but also assists with intervention evaluation and supports clinical treatment approach with MCID outcomes (Bloom et al., 2021; Hanff et al., 2024). Using these methods, researchers and clinicians will be able to derive more precise and patient-specific MCIDs that correlate more accurately with PROs.

The complicated nature of some PROs and differences in patient characteristics is not well dealt with using one variable or one factor or overly simplified methods most commonly used in the standard MCID calculation approaches. These shortcomings are solved by MVA, SEM, and ML approaches.

This SEM approach has been a validated technique that has been used to test the conceptual models that underpin PROMs in MCID studies. Let us say that there are latent concepts in QoL that MVA influences and alters what MVA entails. SEM in this manner has been applied to test the outline that the clinical changes prompt specific observable outcomes. It has employed SEM to show that concepts like the QoL form a bridge between the clinical interventions and observable outcomes. SEM has been used, for example, to test how latent dimensions like QoL moderate the impact of clinical changes on direct observable outcomes (Terwee et al., 2007). Terwee et al. (2007) emphasize the need for quality criteria in the case of QoLs. However, King (2011) points out that there is no standardized MCID.

Meta-Analytic Approaches to Synthesizing MCID Data

Meta-analytic methods improve the accuracy of MCIDs across various patient demographics and clinical conditions; therefore, we need meta-analyses like in other nonclinical fields. One example is the work of Mouelhi et al. (2020), who investigated the varying anchors and methods used in the estimation of HRQoL instruments MCIDs. That difference further exposes the gap for meta-analysis. Importantly, the work by Johnston et al. (2015) presented criteria for judging the credibility of the MCID estimates and the importance of methodology in a meta-analytic synthesis.

Compiling SDs from several research studies facilitates the computation of pooled SDs, which can be multiplied by a selected factor (often 0.5) to approximate a consolidated MCID. Watt et al. (2021) illustrated this by aggregating SDs from randomized controlled trials (RCTs) to establish MCIDs for cognitive assessment instruments, thus ensuring that the derived MCID represents a wider patient demographic. Also, integrating data from studies employing PROs or clinical anchors to ascertain MCIDs facilitates the amalgamation of patient-centered criteria.

In addition, the factors such as differences in statistical methods and anchors used in the different analyses could lead to conflicting formulations of MCID and therefore affect the synthesis works. Variations in the demographic or clinical characteristics across the studies may limit the applicability of the aggregated MCID estimates.

The use of *meta-analytic techniques* toward the synthesis of MCID estimates facilitates the integration of different estimates, which results in these parameters being more useful in clinical practice. A thorough systematic review of existing studies on the topic, along with the gathering of the data through distribution- and anchor-based methods techniques, enables researchers to calculate clinical significance.

In sum, the formulation of MCID and MID using different methodological approaches greatly impacts the decision-making process in a data-driven health system compared to formulation done using uniform methods. Distribution-based, anchor-based, or hybrid approaches all provide added value toward the understanding of clinically important differences (e.g., Copay et al., 2007; Mouelhi et al., 2020). These techniques enhance the care that health-care providers give to their patients, determine the treatment strategies that will be employed, and enhance the health outcomes.

References

Aarons, G. A., Sklar, M., Mustanski, B., Benbow, N., & Brown, C. H. (2017). "Scaling-out" evidence-based interventions to new populations or new health care delivery systems. *Implementation Science: IS, 12*(1), 111. https://doi.org/10.1186/s13012-017-0640-6

Bloom, J. R., et al. (2021). The minimal clinically important difference: A review of clinical significance. *The American Journal of Sports Medicine, 51*(2), 1683–1690. https://doi.org/10.1177/03635465211053869

Byun, J., Lai, D., Luo, S., Risser, J., Tung, B., & Hardy, R. J. (2013). A hybrid method in combining treatment effects from matched and unmatched studies. *Statistics in Medicine, 32*(28), 4924–4937. https://doi.org/10.1002/sim.5887

Copay, A. G., Subach, B. R., Glassman, S. D., Polly, D. W., Jr., & Schuler, T. C. (2007). Understanding the minimum clinically important difference: A review of concepts and methods. *The spine Journal: Official Journal of the North American Spine Society, 7*(5), 541–546. https://doi.org/10.1016/j.spinee.2007.01.008

Gupta, A., Bhint, S., Rajak, S. K., & Majumdar, M. (2023). Determining minimal clinically important differences in the north star ambulatory assessment (NSAA) for patients with Duchenne muscular dystrophy. *PLoS One, 18*(3), e0283669. https://doi.org/10.1371/journal.pone.0283669

Hanff, A.-M., Krüger, R., McCrum, C., & Ley, C. (2024). Mixed effects models but not t-tests or linear regression detect progression of apathy in Parkinson's disease over time: A comparative analysis. *BMC Medical Research Methodology, 24*, 183. https://doi.org/10.17605/OSF.IO/NF4YB

IQWiG. (2023). *General methods version 7.0*, https://www.iqwig.de/methoden/general-methods_version-7-0.pdf

Johnston, B. C., Ebrahim, S., Carrasco-Labra, A., Furukawa, T. A., Patrick, D. L., Crawford, M. W., Hemmelgarn, B. R., Schünemann, H. J., Guyatt, G. H., & Nesrallah, G. (2015). Minimally important difference estimates and methods: A protocol. *BMJ Open, 5*(10), e007953. https://doi.org/10.1136/bmjopen-2015-007953

Kamper, S. J., Maher, C. G., & Mackay, G. (2009). Global rating of change scales: A review of strengths and weaknesses and considerations for design. *The Journal of Manual & Manipulative Therapy, 17*(3), 163–170. https://doi.org/10.1179/jmt.2009.17.3.163

King, M. T. (2011). A point of minimal important difference (MID): A critique of terminology and methods. *Expert Review of Pharmacoeconomics & Outcomes Research, 11*(2), 171–184. https://doi.org/10.1586/erp.11.9

Mouelhi, Y., Jouve, E., Castelli, C., & Gentile, S. (2020). How is the minimal clinically important difference established in health-related quality of life instruments? Review of anchors and methods. *Health and Quality of Life Outcomes, 18*(1), 136. https://doi.org/10.1186/s12955-020-01344-w

Peeters, M., et al. (2023). Towards biomarker-based optimization of deep brain stimulation in Parkinson's disease patients. *Frontiers in Neuroscience, 17*, 1091781. https://doi.org/10.3389/fnins.2022.1091781

Pinheiro, J., Bates, D., Deb Roy, S., Sarkar, D., & R Core Team. (2018). Nlme: Linear and nonlinear mixed effects models. *R Package Version, 3*, 1–137. http://CRAN.R-project.org/package=nlme

Schober, P., & Vetter, T. R. (2021). Linear mixed-effects models in medical research. *Anesthesia & Analgesia, 132*(6), 689–696. https://doi.org/10.1213/ANE.0000000000005541

Sobreira, R. A., Pereira, L. F. F., Santos, R. S., & Almeida, L. F. W. (2020). Minimal clinically important differences for measures of pain, lung function, fatigue, and functionality in spinal cord injury. *Physical Therapy, 100*(4), 639–653. https://doi.org/10.1093/ptj/pzaa210

Terwee, C. B., Dekker, F. W., Wiersinga, W. M., Prummel, M. F., & Bossuyt, P. M. (2007). On assessing responsiveness of health-related quality of life instruments: Guidelines for instrument evaluation. *Quality of Life Research, 12*(4), 349–362. https://doi.org/10.1023/A:1023499322593

Watt, J. A., Veroniki, A. A., Tricco, A. C., & Straus, S. E. (2021). Using a distribution-based approach and systematic review methods to derive minimum clinically important differences. *BMC Medical Research Methodology, 21*(1), 41. https://doi.org/10.1186/s12874-021-01228-7

Wood, S. N. (2017). *Generalized additive models: An introduction with R* (2nd ed.). Chapman and Hall/CRC. https://doi.org/10.1201/9781315370279

Wu, J., Liu, Y., Zhang, L., & Liu, Y. (2020). Establishing minimal clinically important differences for the quality of life instrument in patients with esophageal cancer QLICP-ES (V2.0) based on anchor-based and distribution-based methods. *Research Square., 12*, 753729. https://doi.org/10.21203/rs.3.rs-57319/v1

Zhang, D. Y., Chen, S. X., & Yin, P. (2012). Optimizing the specificity of nucleic acid hybridization. *Nature Chemistry, 4*(3), 208–214. https://doi.org/10.1038/nchem.1246

Zhu, Y., Wei, H., Wang, Y., & Zhang, J. (2023). Minimum clinically important differences in the Minnesota living with heart failure questionnaire: From a study of heart failure patients treated with integrated Chinese and Western medicine. *Frontiers in Cardiovascular Medicine, 10*, 1242216. https://doi.org/10.3389/fcvm.2023.1242216

Chapter 6
MCID/MID in Health-Technology Assessment (HTA) in Different Countries: Controversies

Abstract The evaluation of minimal clinically important difference (MCID) and minimal important difference (MID) has become central to regulatory and policy-making in health technology assessment (HTA). MCID refers to the smallest change in a patient-related outcome that a patient would identify as noticeable or perceivable. While institutions like the FDA and EMA increasingly emphasize MCID in evaluating patient-reported outcomes (PROs), challenges persist due to its inconsistent application and lack of standardization across countries and HTA bodies. For instance, Germany's Institute for Quality and Efficiency in Health Care (IQWiG) employs a fixed ≥15% improvement threshold, but this has been contested due to variability in diseases, scales, and patient profiles.

MCID plays a crucial role in both clinical trials and postmarket surveillance, as it helps determine whether observed improvements yield meaningful benefits for patients or not. However, a universal threshold has been criticized by many because of significant variations in baseline characteristics such as age, disease severity, comorbidities, treatment history, ethnicity, sociocultural status, or clinical context, among others. Consequently, some argue for an individualized approach that tailors MCID to specific conditions and populations. This method would consider factors like ethnicity, drug tolerance, disease progression over time, or even culture.

Across global HTA systems, MCID's implementation diverges. Agencies such as NICE in the United Kingdom and HAS in France differ significantly from IQWiG in Germany in their reliance on MCID. The use of real-world evidence (RWE), particularly in rare diseases and gene therapies, is emerging as a way to validate clinical trial results and complement MCID assessments. However, even within Europe, varying methodological standards and thresholds often lead to different reimbursement and approval outcomes for the same drug, as seen in the contrasting evaluations.

The lack of consensus on how to calculate MCID—whether through anchor-based, distribution-based, or hybrid methods—further complicates its utility. Studies show inconsistent usage of MCID across specialties and tools, reflecting the absence of a robust methodological framework. This inconsistency impedes the comparability of research findings and policy decisions, especially in

I. Fekete, *Transforming Data into Informed Decisions across Clinical and Non-Clinical Domains*, https://doi.org/10.1007/978-3-032-02595-1_6

cost-effectiveness analysis (CEA), where MCID can clarify whether incremental clinical benefits justify additional costs.

Recent regulatory efforts within the European Union aim to harmonize HTA procedures through initiatives like the EU HTA Regulation and Joint Clinical Assessments (JCA). These aim to reduce redundancy and facilitate timely access to innovative therapies. Still, disparities persist due to differences in political will, health-care infrastructure, and economic capacity, particularly in low- and middle-income countries. Therefore, while standardization could enhance comparability and meta-analyses, a flexible, patient-centered MCID approach remains crucial for accurate and equitable health-care decision-making.

Keywords Health-technology assessment (HTA) · Regulatory perspectives on HTA · MCID/MID application in HTA across countries · Cost-effectiveness analysis · European Health Technology Assessment (EU HTA) · Joint EU HTA · EUnetHTA · EMA · FDA · Controversies in HTA-decisions

The Importance of MCID in Regulatory Approvals and Health-Care Policies

When evaluating novel medications, medical devices, or treatments, regulatory agencies like the European Medicines Agency (EMA) and the US Food and Drug Administration (FDA) are in charge. I have just named two organizations because they are world known.

Although primary and secondary clinical endpoint data are frequently included in regulatory submissions (in German, they call them "Dossiers"), patient-perceived benefits/perceptions may not always be correlated clinically with these patient-reported outcomes (PROs) [e.g., Institute for Quality and Efficiency in Health Care (IQWiG), 2023]. A statistically significant decrease in symptoms could not be enough for chronic illnesses like rheumatoid arthritis (RA) until the improvement reaches above the ≥15% minimal clinically important difference (MCID) threshold (IQWiG, 2023, p. 173). Benefit–risk assessment is particularly important for treatments like *biologics* or *chemotherapeutic drugs* that (might) have serious adverse effects.

By determining whether actual results match clinical trial assumptions, MCID can direct postmarket surveillance after initial approvals, not just in Phase 3. Hence, additional regulatory review may be necessary for treatments that do not fulfil MCID thresholds in a variety of demographics.

MCID is being used more and more frequently in health technology assessment (HTA) to establish coverage and reimbursement guidelines. However, regarding standardization (i.e., a universal MCID), I have brought up several arguments against it such as the variance in patient baseline characteristics (demographic characteristics such as age, personal preferences, comorbidities, concomitant medications, prior treatment, particular disease circumstances, therapies, progression rate, therapy type, follow-up duration, disease dynamics, patient subjective perceptions, therapy responses or drug tolerance cultural factors, ethnicity, or socio-economic status).

I should think, most importantly, the clinical indication itself. I argue that a uniform MCID cannot capture all clinical indications via a universal threshold given that the progression and severity of these diseases show great diversity.

First, indication-specific MCID values can differ among population and clinical conditions. Second, robust methodological approaches are necessary for establishing MCID thresholds, frequently combining distribution- and anchor-based procedures. Their broad adoption may be hampered by the absence of established procedures. Policy decisions also need to take cost-effectiveness [cost-effectiveness analysis (CEA)], safety, and feasibility into account. There is one argument, which I accept and advocate: standardization is the foundation stone of reliability and comparability of MCID across studies. Otherwise, no meta-analysis can be performed whatsoever.

This chapter is about HTA decisions in various countries and controversies around it (for an overview on HTA in general, see, e.g., Fontrier et al., 2022). When it comes to HTA, what comes to mind are usually HTA agencies such as the FDA in the United States, EMA in Europe, Canadian Agency for Drug and Technology in Health (CADTH), Haute Autorité de santé (HAS) in France, IQWiG in Germany, or National Institute for Health Care Excellence (NICE) in England, to name a few. These agencies differ immensely regarding their protocols, for example, as to whether they accept real-world data (Graili et al., 2023) or whether (and, if yes, how) they use MCIDs (Larose et al., 2023), or whether they use traditional *p*-value testing (IQWiG, 2023).

Real-world data (RWD) or real-world evidence (RWE) are collected outside laboratory settings. Registries or electronic health records (EHRs) are good examples of RWD. For statistical problems and solutions about registries, see, for example, Fekete et al. (2020).

Importantly, HTA in Europe has changed a lot. Namely, there is not a future cooperation between EU countries, which came into effect in January 2025. Every country will fall under HTA Regulation (HTAR). As of February 2025, there are 22 guidelines and 19 templates that these countries will have to follow.[1] For further information on this harmonization process, consult Urbina et al. (2024).

Because this book is on MCID/minimal important difference (MID), I do not mean to deviate from this metric, for example, by writing about incorporating RWE or *p*-values in HTA decisions. I would like to focus on MCID/MID in the context of HTA only. HTA decides upon drug approval, reimbursement, and the implementation of new medical technologies and drugs everywhere in the World (in Germany, e.g., it is the G-BA[2]). The G-BA document in Germany (which is regulated by the

---

[1] https://www.iqwig.de/en/presse/press-releases/press-releases-detailpage_133632.html

[2] https://www.g-ba.de/english/

law §35a of the SGB V.[3]), for example, outlines the criteria for early benefit assessment of pharmaceuticals (one criterion is, e.g., MCID), including the thresholds for determining added benefits in PROs.

In this chapter, I present the uncertainty and debate around the challenges of standardization of MCID in HTA. Some HTA bodies might not use MCID at all (such as the HTA body NICE in the United Kingdom and the HTA body HAS in France; status January 2025), relying instead on alternative measures or thresholds to evaluate the clinical significance of medical interventions (Larose et al., 2023).

For example, the German institute IQWiG[4] (The Institute for Quality and Efficiency in Health Care; in German: Institut für Qualität und Wirtschaftlichkeit im Gesundheitswesen, IQWiG), adheres to a 15% improvement (or decrease) as MCID, as of now (IQWiG, 2023). That is, there should be at least 15% increase (or decrease) on a scale, which is used as MCID on scales as a measure of a noticeable, beneficial, meaningful, and patient-relevant change in HRQoLs such as QoL (IQWiG, 2023, p. 173). For me, it is not conceivable how 15% can be set as an MCID on an interval scale, as we do not just have Likert scales when assessing HRQoLs. However, besides the official IQWiG document (IQWiG, 2023, p. 173), this metric was already defended in 2021 in a press release.[5]

This inconsistency across countries (see, e.g., Larose et al., 2023) does impact the comparability of HTAs across different jurisdictions, potentially affecting patient access to treatments regarded as beneficial in one country but not in another. The challenge is to harmonize these approaches to MCID to ensure efficient healthcare delivery around the globe and, last but not least, to be able to compare clinical trials and patient improvement.

No doubt that MCID/MID should be a central tool for evaluating clinical effectiveness (and also in the evaluation of potential harm), yet it has not been implemented and used globally (e.g., Larose et al., 2023). The fact that MCID/MID varies immensely across different countries (Larose et al., 2023) entails that there a non-standardized MCID can lead to inconsistencies in the *(mis-)interpretation of clinical evidence* and *comparability of clinical studies*, but also in *clinical decision-making* (e.g., about the continuation, discontinuation, or modification of a therapy or the evaluation of patient-relevant improvement). Ultimately, these inconsistencies lead to inconsistencies in patient care.

This chapter discusses these topics around MCID/MID in HTA across different countries, particularly, in Germany, the United Kingdom, and France (for a comparison of the three countries, see, e.g., Larose et al., 2023). I would like to highlight the debate of variability across countries without taking a position between the two sides: (i) harmonization/standardization (a universal/singular MCID) versus (ii) indication-specificity and flexibility.

---

[3] In English: §35a Social Code Book [SGB] V.

[4] https://www.iqwig.de/en/

[5] https://www.iqwig.de/en/presse/press-releases/press-releases-detailpage_27520.html

The latter approach (ii) is very close to the concept of individualized (or personalized) medicine. However, admittedly, a consistent evaluation of health-care interventions and medical products globally would ensure comparability of medical products and interventions.

Regulatory Perspectives on MCID (FDA, EMA Guidelines)

The significance of assessing outcomes that represent patients' experiences is emphasized in the FDA's PRO advice (FDA, 2009). Determining whether patients find improvements in PROs measures—like pain, exhaustion, or functional status—meaningful is largely dependent on MCID. For instance, patients show improvements on the MCID in the development of medicines for chronic illnesses for these medicines to be considered as *clinically relevant* (FDA, 2009).

Benefit–risk assessment is crucial in that the benefits should outweigh the potential dangers, the latter also called "adverse effects" during the therapy in the pharmaceutical world. In this context, MCID serves a crucial role. This is especially important for treatments that have major adverse effects, like *gene therapies* or *cancer medications*.

RWE has been accepted by the FDA to supplement clinical trial data. The idea behind this is that the FDA wants to make sure that real-world therapy effects match patient expectations and to speed up the approval process. RWE, for example, can confirm if a novel intervention surpasses its MCID in a variety of demographic groups.

The need for relevant measurements that represent patient priorities is emphasized in the EMA's guidance on endpoints for clinical studies. To assess if primary and secondary endpoints reflect changes that are clinically significant, MCID offers a standard. For instance, EMA commonly uses MCID levels to specify acceptable efficacy standards in *respiratory medicine* and *rheumatology* (personal communication with medical experts). But what about the rest of the diseases?

While both organizations (FDA and EMA) recognize MCID, their regulatory approaches diverge. The FDA and EMA prioritize PROs; nevertheless, the EMA has provided more comprehensive guidance on including MCID in PRO development and validation. The EMA's flexible approach offers increased flexibility in the application of MCID for rare diseases such as spinal muscular atrophy (SMA). This, in turn, facilitates early access to medicines based on preliminary results (European Medicines Agency, 2021). For example, the EMA was flexible toward SMA, a rare and life-threatening disease, currently treated by nusinersen (Mercuri et al., 2018).

In clinical trials for Spinraza® (nusinersen) by the pharmaceutical company Biogen, researchers used measures like the Hammersmith functional motor scale-expanded (HFMSE) as an MCID. However, the EMA allowed some degree of flexibility when interpreting these results, because SMA is a rare disease with limited treatment options.

The EMA granted Spinraza® a so-called conditional marketing authorization (they call it "facilitated approval"). This decision acknowledged the rarity of the disease. The fear was that a strict application of MCIDs might have delayed patient access (European Medicines Agency, 2021).

By the way, speaking of rare diseases (orphan drugs), it happens very rarely, but Germany has already allowed RWE collection. They did it for the gene therapy of Zolgensma®, by the pharmaceutical company Novartis (brand name for a drug to treat SMA).[6] They add that more mandates for RWE may come in the case of gene therapies. This was the first time in Germany that the authority allowed the collection of RWE.[7]

Similarly, the FDA employs such latitude in its accelerated approval and breakthrough therapy initiatives. It is also crucial to incorporate RWE and MCID assessments in postmarket phase due to generalizability (ecological validity).

Notwithstanding its advantages, there are disadvantages to employing MCID in regulatory processes. The lack of established methods for its calculation complicates the consistent application of MCID across trials and populations. Due to the variability in MCID levels based on population, condition, or outcome measure, regulators must adopt a flexible approach in their interpretation of MCID.

Lastly, *digital biomarkers* and *wearable technology* provide new ways to record PROs, which could track and improve MCID's accuracy. In my opinion, streamlining regulatory procedures for international research may be possible by coordinating the FDA and EMA's approaches to MCID through cooperative efforts.

Variability across HTA (Health Technology Assessment) Bodies in the World

> There were concerns that a strict 15-percent threshold for responder analyses would not be applicable to all scales, making it more difficult to prove added benefit. These concerns were unfounded. The addenda published today show that the new threshold is feasible in practice."[8]
>
> (Katrin Nink,[9]
> IQWiG-Ressort ArzneimittelbewertungCologne, Germany, February 2021)

[6] https://www.rjwpartners.com/post/germany-mandates-real-world-data-collection-for-gene-therapy-zolgensma

[7] https://www.rjwpartners.com/post/germany-mandates-real-world-data-collection-for-gene-therapy-zolgensma

[8] "Clinical relevance of patient-reported outcomes: new threshold proven feasible in practice," https://www.iqwig.de/en/presse/press-releases/press-releases-detailpage_27520.html, February 18, 2021

[9] Katrin Nink is currently Head of Division Oncology (Solid Tumours) and Routine Practice Data Collection, Drug Assessment Department, Institute for Quality and Efficiency in Health Care (IQWiG) in Cologne.

> The wide-ranging thresholds obtained with the different methodologies make it difficult to evaluate the real effectiveness of a given treatment, questioning the usefulness of the MCID, as currently available, in the clinical research.[10]
>
> Marco Franceschini et al. (2023), Researchers, Investigation performed at Clinica Ortopedica e Traumatologica 2, IRCCS Istituto Ortopedico Rizzoli, Bologna, Italy, 2023

I would like to begin by presenting the two citations above, which elucidate the tension between two "camps": one who defend a standardized and universal metric, and the second one who question the usefulness of MCID (Franceschini et al., 2023).

From IQWiG, Nink is from the Drug Assessment Department and is focused on the perspective of standardization. Katrin Nink's viewpoint (IQWiG), as shared in a press release in February 2021, seemed to overwhelmingly affirm the ≥15% threshold on a bounded scale. However, this raises concerns on its relevance to other scales, such as an interval scale.

The German Federal Joint Committee (G-BA; Gemeinsamer Bundesausschuss), one of the major players in the EUnetHTA21 (European Network for Health Technology Assessment joint consortium for future EU HTA regulation) methodological workstream, made this important methodological change regarding the assessment of the added benefit of drug interventions.

This change includes the creation of a global single response threshold (i.e., responsiveness to some treatment), defined as a minimum of a ≥ 15% relative improvement within the scale range of the measurement instrument on all patient outcomes reported by the respondents, regardless of the medical condition.

According to Katrin Nink, the practical application of the ≥15% IQWiG threshold in assessments of drugs like Secukinumab, Ivacaftor, and Alpelisib proved its efficacy. Another example for the application of the ≥15% threshold is the clinical study report (dossier) for drug assessment and marketing authorization of dimethyl fumarate[11] (Tecfidera™) of the pharmaceutical company Biogen GmbH in Germany.[12]

Even though its significance in evidence-based health-care decision-making is widely acknowledged, HTA bodies' operations vary greatly across the globe. This variation results from variations in each nation's or region's health-care systems, legal frameworks, medical expert opinions, economic circumstances, cultural backgrounds, demographics, etc.

---

[10] Marco Franceschini et al. (2023), "The Minimal Clinically Important Difference Changes Greatly Based on the Different Calculation Methods", https://journals.sagepub.com/doi/pdf/10.1177/03635465231152484, 2023

[11] Dimethyl fumarate is for the treatment of adult patients with relapsing-remitting multiple sclerosis.

[12] https://www.g-ba.de/downloads/92-975-6860/2022_06_08_Modul4B_DMF.pdf, June 10th, 2022

The governing mechanisms of HTA bodies vary. HTA may be performed by independent organizations or academic institutions in certain nations, such as the US's Institute for Clinical and Economic Review [ICER], or by government-affiliated agencies, such as the UK's National Institute for Health and Care Excellence [NICE]).

HTA bodies have divergent objectives depending on the region. In low- and middle-income countries, there is an emphasis on the price and availability of vital drugs, whereas in more developed countries the concentration is on cost-efficiency and value. Thailand's ICER sets a standard for cost-effectiveness of new treatments. However, Thailand's Health Intervention and Technology Assessment Program (HITAP) is more focused on providing equal opportunity for health-care innovations.

Given the various rules and regulations, it is understandable why HTA bodies have varying methods of evaluation. For example, cost-effectiveness analysis is frequently adopted by most HTA bodies. However, the standard of measuring cost-effectiveness differs. To illustrate, NICE sets the criteria between £20,000 and £30,000 for each quality-adjusted life year. However, WHO argues that the standards are GDP per capita for low- and middle-income countries.

Caregiver burden and productivity losses are among the wider societal perspectives taken by some HTA organizations like NICE, whereas other organizations, such as Germany's Institute for Quality and Efficiency in Health Care (IQWiG, 2023) prioritizes clinical efficacy, considering evaluation regarding QoL and long-term outcomes with direct costs.

Disparities in the quality and availability of local data in low- and middle-income countries (LMICs) may give rise to reliance on foreign data, which may not always be indicative of domestic conditions due to differences in ethnicity or genetics, for example.

A one-size-fits-all metric such as the ≥15% threshold in Germany (IQWiG, 2023) is unworkable, in my opinion, since different nations place varying priorities and have varying legal context. Second, the adoption of advanced HTA approaches is hampered in LMICs by a lack of funding and technical know-how. HTA methods and their execution are also influenced by political will and cultural perspectives on health-care spending. In my view, another problem could be that medical doctors might not reach a conclusion on a one-size-fits-all metric or an MCID for a single medical condition.

There are continuous attempts to standardize HTA procedures worldwide. Knowledge sharing is made by HTA bodies and is encouraged by groups such as Health Technology Assessment International (HTAi) and the International Network of Agencies for Health Technology Assessment (INAHTA). In addition, the WHO's effort to improve HTA in LMICs seeks to promote global cooperation.

A Comparative Analysis of MCID/MID Application in HTA Across Countries

A lack of consensus within the medical profession over the standardization of MCID still persists despite the fact that a lot of people put their faith in medicine. This lack of unanimity is due to dozens of factors including interindividual differences among a lot of others. This lack of homogeneity is not solely theoretical; it also appeared in the realm of HTA, as mentioned in the previous chapter. I would compare it to the jurisdiction of various countries. Because various HTA agencies use unique metrics and thresholds for determining pharmacological efficacy, no wonder the outcome is divergent. This is the reason why we have divergent results for identical pharmaceuticals and clinical data across several countries.

The standardization of MCID is a challenge in clinical and psychological research (but also in every field, of course). Although standardization seeks to establish *uniformity* and *comparability* among research findings and treatments, it faces numerous obstacles and counterarguments. I personally believe that comparability is the only reason that speaks for standardization. I believe MCID/MID should be a dynamic concept. So dynamic that it can even change within a person as the disease progresses or deteriorates.

According to Marco Franceschini et al. (2023), "various MCID calculation methods result in considerable variability in values, which substantially influences the percentages of patients attaining the MCID." The extensive array of MCID approaches outlined in this book underscores the challenge of selecting the optimal approach to ascertain the most suitable value. Regrettably, no strategy proved superior, leaving the MCID for practical practice ambiguous.

These thoughts echo with Marco Franceschini et al. (2023), who claim that "a further contributing factor to the variability in reported MCID scores is the study population." Hence, standardizing the MCID necessitates agreement on the best suitable method. Hence, I advocate for a more personalized approach contra a standardized one.

Marco Franceschini et al. (2023) referred to a study demonstrating that, among individuals with diverse knee impairments, especially women who originally showed a high functional status score and subacute symptoms required smaller score variations to achieve the MCID. In my opinion, this observation can serve as an argument against a universal MCID.

In conclusion, although the standardization of MCID would enhance the comparability of research findings, it would encounter considerable hurdles. Next, we are going to examine various HTA bodies based on (Larose et al., 2023) and the German IQWiG.

The existence of multiple MCIDs/MIDs (Larose et al., 2023) is due to the legal sovereignty of these countries according to Fontrier et al. (2022). Also, differing expert opinions among HTA bodies play a role, as there are national expert panels. In Germany, it is the G-BA (*Gemeinsamer Bundesausschuss* in German; translated as "German Federal Joint Committee" in English), which is responsible for and has mandated the IQWiG for to conduct early benefit drug assessment and the G-BA one of the stakeholders in EUnetHTA21 (European Network for HTA). Hence, G-BA is the highest decision-making authority in Germany in the German healthcare system, which is responsible for determining the benefits covered by the German health insurance companies (for an overview on HTA in general, see, e.g., Fontrier et al., 2022).

However, a very interesting discrepancy emerges as to how various HTA bodies in different countries determine MCID "around the globe" (for a comparison of Germany, France, and the United Kingdom, see, e.g., Larose et al., 2023). This variation basically entails that the same composition (in the sense of medicine) can have different level(s) of efficacy "if you cross the German-French border."

To put it bluntly, the statistical analysis of clinical data undergoes a different statistical analysis process depending on the country, specifically, the protocols defined by the competent authority of the given country (e.g., it is the G-BA and the IQWiG in Germany). HTA bodies in the world govern this procedure. These disparities arise from the absence of a uniquely defined and internationally endorsed metric (such as MCID) for comparison (Larose et al., 2023). I would liken this variability to that observed in legal systems across the globe, for example, case law in England and India compared to continental law in Europe, or even minor divisions within Europe.

It follows from what has been said previously that inconsistencies in how clinical evidence is assessed arise, which, in turn, affect the evaluation of PROs. Ultimately, these inconsistencies influence decision-making based on, or derived from, *exactly the same clinical dataset* (!). In turn, decision-making influences both approval (early benefit assessment) and reimbursement of new medical interventions or compositions/medicines (IQWiG, 2023).

An example of two controversial HTA decisions involving two countries that led to different conclusions based on exactly the same clinical data marketed under the composition name of ruxolitinib (Jakavi® by Novartis Pharma GmbH or Jakafi®[13]). This drug is used to treat myelofibrosis (MF) (https://www.jakafi.com/). Ruxolitinib was assessed by multiple HTA bodies. The company could demonstrate a significant reduction in spleen size and improvement in symptoms for patients with MF.

The German authority in 2014 evaluated ruxolitinib and gave it a low rating. At that time, they already mentioned MID in Module 4 in their dossier, see page 132.[14] For the English version of the dossier, see the English version of the IQWiG document.[15]

Ruxolitinib was assessed for both its clinical efficacy and cost-effectiveness derived from the COMFORT-I and COMFORT-II studies, which demonstrated a significant reduction in spleen size and improvement in symptoms for patients with

---

[13] https://www.jakafi.com/

[14] https://www.g-ba.de/downloads/92-975-515/2014-05-14_Modul4A_Ruxolitinib.pdf, see page 132 of Module 4.

[15] https://www.iqwig.de/download/a15-35_ruxolitinib_addendum-to-commission-a15-13.pdf

Myelofibrosis (MF). Ruxolitinib proved to be the best therapy for MF (Harrison et al., 2012).

However, in Germany, the IQWiG and the Federal Joint Committee (G-BA) conducted the early benefit assessment, but despite the positive clinical results from the COMFORT trials, IQWiG concluded that the added benefit of ruxolitinib was not proven for certain subgroups of patients (e.g., those who were ineligible for bone marrow transplantation) and there was also problem with the sample size (see pages 10–12, 14, 15, 54 of Module 4A of the IQWiG dossier). Hence, the G-BA ultimately ruled that the drug offered only a marginal additional benefit (see pages 7 and 63 of the IQWiG dossier).

In contrast, NICE[16] in the United Kingdom came to a different conclusion based on the same clinical data (!). The benefits in the COMFORT studies were acknowledged by NICE. Such differences in the assessments of the same composition impacted the approval and reimbursement processes of ruxolitinib in the two countries. This ruxolitinib example illustrates how two HTA agencies can come to different results based on identical clinical data. What is important for us is that there is no clear threshold for MCID in NICE (Larose et al., 2023).

In the United Kingdom, NICE's recommendation facilitated broader access to the drug, though still with some restrictions. The drug was more widely reimbursed within the National Health Service (NHS), which helped ruxolitinib achieve better market access. I am just mentioning these differences in market access and reimbursement in order to highlight the implications of utilizing MCID or not. Employing MCID can affect *pricing* and *reimbursement negotiations*, the business perspective of MID/MCID. This is why a whole chapter is devoted to the business aspect. Obviously, the bigger the improvement (MCID surpassed), the stronger the case; hence, a higher price can be demanded for the same drug.

The question arises as to what exactly caused the discrepancy between the two assessments? Germany's IQWiG placed a strong emphasis on subgroup analysis and the lack of evidence for specific patient populations. In contrast, NICE in the United Kingdom has established cost-effectiveness thresholds that allow for a more balanced consideration of clinical benefits and costs.

This instance involving ruxolitinib illustrates how two HTA organizations can arrive at varying conclusions with the same clinical data owing to differing frameworks, which eventually impact the approval and reimbursement of new medical therapies and products.

Also, in France and the United Kingdom (NICE) clear and standardized thresholds for clinical relevance (i.e., MCID) have not been established (Larose et al., 2023), in contrast to the standardized ≥15% threshold used in Germany proposed by the IQWiG (IQWiG, 2023, p. 173).[17] Larose et al. (2023) demonstrated in a

[16] NICE (National Institute for Health and Care Excellence) is the HTA body in the United Kingdom.

[17] For the English version, consult https://www.iqwig.de/methoden/general-methods_version-7-0.pdf

systematic review that the French HTA[18] is "more restrictive in its approach than IQWiG." In my mind, this deviation has implications for the generalizability of research findings consistent with Larose et al. (2023).

I personally favor a patient-centric and individualized approach in MCIDs and MIDs *independently from scientific field* in contrast to the German IQWiG system as it is now. I have an approach in mind, whereby baseline characteristics of the patients are considered (e.g., gender, age, ethnicity, BMI, smoking status, medical history, disease severity, baseline characteristics, concomitant medication, laboratory values, and comorbidities depending on the disease). This way, we can tailor MCIDs/MIDs more efficiently to the patient, I should think. For such an approach, we need *subgroup stratification* and advanced statistical modeling which considers *disease progression* and the *status of the patient over time*.

Methodological Diversities in Different HTA Systems

MCID/MID is determined by numerous factors, including the diversity of clinical conditions and the specific disease (Cook, 2008), subjective patient experiences (Engel et al., 2018; Cook, 2008), disease severity, patient heterogeneity (Franceschini et al., 2023), study population and demographic factors (Franceschini et al., 2023), variations in measurement tools, treatments (Franceschini et al., 2023), clinical expertise, ongoing research developments and different MCID calculation methods (Franceschini et al., 2023), differing HTA regulatory frameworks "around the globe" (Larose et al., 2023; Fontier et al., 2022), and even cultural influences that all affect what is considered a *meaningful clinical change*. To illustrate, some people might tolerate pain more (subjectively) than others, so their MCID might be lower.

The abovementioned factors hinder harmonization/standardization efforts of MCID around the globe. Second, to my mind, standardizing MCID would require consensus on which method is the most suitable one, which is challenging from a methodological-statistical perspective.

Regulatory and Policy Frameworks Influencing MCID/MID in HTA

Fig. 6.1 exhibits the differences among three countries, France, the United Kingdom, and Germany.

The existence of various metrics for the concepts of MCID and MID, along with discrepancies among different HTA bodies, stems from the differing expert viewpoints among these HTA organizations. MCID/MID is affected by various clinical

[18] Haute Autorité de Santé, or HAS.

	France	UK	Germany
Overall	While HRQoL data is clearly integrated in the Doctrine, the methodological standards expected are as high as for efficacy data & other endpoints	High expectation for QoL data, and a clear preference for generic instruments.	HRQoL data is integral and ranked equally important as morbidity and mortality data.
Open-Label	✗	✓	✓
Endpoint hierarchy	Must be alpha controlled	Not required: NICE uses a different methodology	Does not need to be alpha controlled
Real-world evidence	Can support re-evaluation	✓	✗
Generic vs specific instruments	Disease-specific	Generic (EQ-5D)	Disease-specific
Instrument Validation	Should be clinically validated	Should be clinically validated	Should be clinically validated
Minimum clinically important difference (MCID)	No clear threshold	No clear threshold	MCID of 15% required, with high hurdle to accept data with <15%
Missing Data Threshold	No clear threshold	No clear threshold	30%

Fig. 6.1 Differences between three countries in HTA across some critical aspects. (Source: Hugo Larose et al., "Opportunities to improve the adoption of health-related quality of life evidence as part of the French Health Technology Assessment process," https://www.ncbi.nlm.nih.gov/pmc/articles/PMC10729510/, December 19, 2024)

conditions, patient diversity, clinical proficiency, differing HTA regulatory frameworks, and also cultural influences, to name a few, that shape the definition of MCID (Larose et al., 2023).

Global HTA Harmonization Efforts

I have already touched upon the issue of harmonization/standardization efforts in this chapter. Most recently, an individualized approach has been advocated in determining the MCID (e.g., Schlichting et al., 2022), suggesting that it should be tailored to each patient's unique circumstances, subjective perceptions, and medical condition, rather than relying on a universal standard like the ≥15% MCID proposed by IQWiG (IQWiG, 2023).

In the words of Schlichting et al. (2022), "the ≥ 15% response threshold is incongruent with previously defined and scientifically established thresholds and is not well-suited for universal implementation. Further scientific evidence and discussion among all stakeholders are needed, especially should this universal rule be considered in the context of future joint clinical assessments of health technologies in the European Union scheduled from 2025 onward" (Schlichting et al., 2022).

In the same vein, Franceschini et al. (2023) point out that those women who have a high baseline functional status require greater changes in MCID scores. At the same time, subacute symptoms required smaller score differences (i.e., their MCID is smaller). To cite Franceschini et al. (2023) precisely, "the lack of context-specific values for many populations and treatments, as well as the great heterogeneity of

MCID calculation methods, limits the validity of MCID use. The very wide range of values that can be obtained and used implies the risk of drawing different conclusions in different studies on the same topic" (Franceschini et al., 2023).

Similarly, Cook (2008) critiques the current use of MCID due to its *subjective nature* and *potential for misinterpretation in clinical practice*. Specifically, Cook (2008) argues that MCID is sometimes treated as a strict threshold rather than a flexible one. He advocates a more refined MCID, which recognizes the *variability in patient perceptions* and the *variability of medical conditions*. De Vet et al. (2023) and Engel et al. (2018), like Cook (2008), also argue along the same line in determining MCID/MID and underscore the *variability in patient subjective perceptions* and the subjectivity of measuring PROs.

To resolve the discrepancy between the defendants of a singular MCID and the other camp (MCID should be personalized), the IQWiG (2023) resolves this debate in a very clever way: when there is an accepted MCID for the given disease (which can be obtained from a medical paper for the given disease), then the institute (IQWiG) accepts that as the MCID. The use of this MCID can also be reinforced by the authority in an official hearing before starting the analysis of the clinical data IQWiG (2023).

Another interesting case in the context of IQWiG is when did not use any thresholds way back it was introduced in their methods paper. In the case of cholinesterase inhibitors in Alzheimer-type disease (final report; Commission No. A05-19A, Version 1.0, 07.02.2007), the IQWiG argued later that this threshold was unhelpful due to high interindividual variation, consistent with Schlichting et al. (2022) who claim that "the≥ 15% response threshold is incongruent with previously defined and scientifically established thresholds and is not well-suited for universal implementation" (Schlichting et al., 2022). From an industry perspective, SmartStep Consulting (2023) also underscores that this threshold ignores a lot of patient and disease characteristics.

Most recently, Nielsen et al. (2023) investigated the use of PROMs and the calculation of MCID in hand surgery research. To this end, Nielsen et al. (2023) conducted a systematic review of the hand surgery literature that used PROMs. From a total of 4677 articles reviewed, 410 met the criteria of including at least one PROM (Nielsen et al., 2023). Among these, only 148 studies mentioned MCID. Only 14 (sic) of these studies calculated MCID values specific to their clinical populations. The review identified around 35 (!) different PROMs and 95 different MCID values (!) sourced from 65 unique articles. Nielsen et al. (2023) highlight great inconsistency reporting of MCIDs in hand surgery literature. One problem can be that these MCIDs were derived from multiple sources, and another one could be that they were calculated employing different metrics and methods. In my view, it is clear now that standardization of MCID suffers from many obstacles that hinder the comparability and interpretation of research findings. Nielsen et al. (2023) are for a standardized approach to calculating and employing MCID values to PROMs to improve the comparability of studies.

First, individual variability in subjective perceptions is a significant challenge: for one person, pain, for example, may be unbearable, while for another, it may seem trivial. Moreover, in my view, as already mentioned, besides age, gender, even cultural values, ethnicity, other demographics, and clinical variables, can all shape MCID.

Regarding the clinical variables, the symptoms and their manifestations, disorders, and psychological conditions and their effects are very divergent varying from individual to individual. To stress again, trying to apply a singular MCID approach seems highly complex in my opinion.

Limitations in research generalizability are also a concern. Findings based on a unified MID may not be applicable or relevant in various contexts or populations. For instance, different medical conditions or stages of illness may require distinct MIDs to capture what constitutes a minimal but meaningful difference in treatment outcomes. In sum, a unified MCID/MID may fail to account for these variations, which might potentially lead to biased or inaccurate assessments in diverse populations. However, I think meta-analyses and systematic reviews can step around the problem of a uniform MCID and the problem of comparability and generalizability.

Lastly, deriving a consolidated MID is a complicated issue with significant methodological and statistical hurdles. For instance, Franceschini et al. (2023) showed that different methods for calculating MCID often produce heterogeneous values which, in turn, greatly influences the proportion of patients meeting the MCID in a certain population. Franceschini et al. (2023) illustratively noted that outcomes of 76% and 24% success for patients in the same series emerged when different MCID values were applied.

Finally, a standardized MID might lead to the neglect of individual patient needs. To exemplify, benzodiazepines, drugs that are prescribed mainly for insomnia, anxiety, acute status epilepticus, panic attacks, among others (e.g., Edinoff et al., 2021) can lead to tolerance. This tolerance (addictive potential) is dependent on age, gender (women are more susceptible to misuse), comorbidities, or the duration of use and abuse (e.g., Edinoff et al., 2021).

In some cases, depending on the duration of use or abuse, individuals do not even feel the effect. So, the question arises: without the knowledge of these abovementioned variables, how can we determine an MCID for, for example, for alprazolam (a benzodiazepine)? The list of drugs is long, where a lot of variables can affect the MCID: sleeping pills, pain-killers, etc. I believe the example with benzodiazepines is a convincing example or just another argument against a standardized, one-size-fit-all MCID metric.

In conclusion, while a unified MCID/MID metric offers simplicity and the potential for standardization, its practical application faces significant challenges. These counterarguments highlight the need for a more personalized approach that considers individual (subjective) patient experiences.

The feasibility and relevance of implementing MID-like metrics in nonclinical fields—such as a "practical or relevant difference"—where direct patient input is usually absent, remain unanswered at this point of the book.

Some open questions include: Why do other disciplines do not have any MIDs? MID is based on statistical and psychometric methods tailored to human subjects research, where individual response variability is both expected and significant. In my mind, MID can and should be extended not just to human subjects but also to disciplines where these measurements are and where they work with statistics.

MCID/MID Considerations and Cost-Effectiveness Analyses (CEA)

By connecting clinical results to economic value, MCIDs can improve the usefulness of economic evaluations when used with cost-effectiveness analyses (CEA). This ensures that interventions offer significant benefits within limited resources. Prioritizing therapies that meet patient-centered benefit thresholds (i.e., MCID) in addition to statistical efficacy is made easier by incorporating MCIDs into CEA.

The incremental cost-effectiveness ratio (ICER) analyses additional expenses per unit of benefit [such as quality-adjusted life year (QALY)]. This is typically assessed by CEAs. Policymakers can ascertain whether the ICER signifies a clinically significant improvement in PROs or just reflects little or no practical significance by incorporating MCIDs into these computations.

MCIDs differ among indications, individuals, populations, or ethnic groups, to name a few. For instance, MCIDs for pain ratings in the treatment of chronic pain may vary based on the tolerance of pain, on comorbidities, or on the demographics of the patient.

CEAs can take into consideration differences in perceived benefit by classifying MCIDs according to patient personalized MCID stratification; or, what I called earlier as an individualized approach. This could be accomplished with the help of sophisticated modeling approaches like generalized additive mixed models (e.g., GAMMs; Wood, 2017).

MCIDs can be easily incorporated into CEA frameworks by mapping them onto utility scales, such as QALY increments. For example, a particular utility gain obtained from patient preference research may be represented by a five-point MCID on a depression scale. Second, involving patients, physicians, and legislators in the determination of MCIDs adds an extra layer of precision in the estimation of MCID. In this context, Delphi methods and expert panels, for example, can be very helpful.

In physical therapy, MCIDs for functional scores—such as walking distance in pulmonary rehabilitation—are essential for assessing the justification of higher spending on medical interventions.

When combined, MCIDs and CEAs provide a strong framework for assessing the effectiveness of interventions. This way, it is guaranteed that financial factors are in line with results that have clinical significance, which is also optimal for stakeholders when it comes to resource allocation decisions, for example.

Decisions on Reimbursement and Comprehensive Allocation of Health-Care Resources

Reimbursement methods constitute an important element of health-care systems. They outline the scope of coverage for medical services, treatments, and technologies. Usually, reimbursement selections are made by a public or private payer, which could be a state, an insurance company, or even a health-care provider.

Cost-effectiveness is commonly articulated as cost per quality-adjusted life year (QALY) or incremental cost-effectiveness ratios (ICERs). How is value incorporated in the operation of a health-care system's budget? Different countries tackle this problem differently; for example, the United Kingdom opts for value-based health care as a priority, in contrast Thailand values access and cost.

My observation is that within the bounds of reimbursement spending, there is a focus on clinically or economically significant PROs. There is often a focus on treatments that significantly improve PROs, whether clinical or economic, within reimbursement spending. Treatment allocators ought to focus on those with a low cost per QALY as it is probable that such an approach will maximize the value of the available resources. Offering reimbursement for expensive new therapies like gene medicine for spinal muscle atrophy (SMA; Novartis has an expensive therapy) would encourage the industry to invest more, but at the same time, it may exhaust health-care funds.

Cost-effectiveness is often at odds with equity and fairness. The small population QALY issue when children are few is especially true for orphan drugs addressing rare diseases such as spinal muscular atrophy (SMA), which leads to very high cost per QALY and has serious effects on how countries are willing to pay for them.

This international cooperation on price negotiations and evidence can improve affordability while reducing duplication of efforts for these international LMICs. NICE's reimbursement mechanism in the United Kingdom remains focused on benefit to society and cost-effectiveness with certain limits on the cost per QALY. While this helps to avoid wastage of NHS resources, it has received negative press for the lack of availability of some challenging expensive treatments.

Medicare has an incomplete fee-for-service system, which reflects the difficulties of this balance in a public-private system. There are inequalities in provision due to decisions being out of sync at the regional level.

Thailand's payment policy under Health Intervention and Technology Assessment Program (HITAP) is focused on cost-effectiveness and equity given the economic and social conditions of the region. This ensures resources are used in line with national priorities.

The allocation of health-care resources in practical terms and the availability, efficiency, and equity of the health-care system are intricately related to reimbursement decisions. While many challenges still exist, some promising solutions offer new hope, including value-based pricing and international cooperation. This can improve HTA decision-making processes.

European Health Technology Assessment (EU HTA)

The HTA method has been very singular in ensuring Europe was evaluating health technologies with respect to their efficacy, safety, and cost-benefit analysis. The reason for having a common EU HTA framework is to cater to the very large gaps which are apparent in the member states concerning the practice of assessment and the availability of new therapies and technologies on the market, as already mentioned above.

As per the regulation effective from January 12, 2025, the joint clinical assessments (JCA) are now going to be easier, which facilitates the use of these assessments more often. Similarly, it encourages the use of JCA to have a more unified approach toward HTA in all of Europe. That, in turn, is supposed to improve the efficiency of the assessment process (Julian et al., 2022a, b).

This regulatory framework is deemed to improve the collaborations between member states, and all this will lead to making the assessments easier (Julian et al., 2022a, b). It is this regulation which highlights the role of stakeholders as key parts of the evaluation process that will, without doubt, affect the transparency and legitimacy of it. This has already been suggested (Van Haesendonck, 2023; Desmet, 2024).

Not strictly related to MCID, but recent developments on the EU HTA (May 2, 2025) are underway. The first Joint Clinical Assessments (JCA) have happened for the innovation of health technologies. This is meant to facilitate the ATMP (Access to Advanced Therapy Medicinal Products) access across Europe for the treatment of melanoma (autologous melanoma-derived tumor infiltrating lymphocytes).

The first JCA is being conducted for lifileucel (Iovance Biotherapeutics, with the indication of advanced melanoma) under the newly introduced EU HTA regulation. Lifileucel, an ATMP in the form of autologous tumor infiltrating lymphocytes (TIL) therapy, was recently granted FDA approval. The JCA is done concurrently with the EMA's marketing authorization procedure. Lead assessor is HAS (France) with AOTMiT (Poland) serving as co-assessor.

Both EMA's validation for marketing authorization application was in March 2025. The Coordination Group of the HTA shall endorse the JCA reports and summary reports not later than 30 days postmarket authorization from the European Commission (EC).

The development of a more unified HTA in Europe already commenced in the early 1990s with several initiatives intended to foster cooperation between the member states. The creation of EUnetHTA in 2006 is considered a major landmark as it set out to establish a sustainable network to facilitate the exchange of HTA information and to build upon best practices (Berntgen et al., 2014; Ruether et al., 2022). All these years EUnetHTA has coordinated several joint assessments and has developed shared methodological frameworks that form the basis of the current EU HTA regulation (Ruether et al., 2022).

In spite of these developments, some serious difficulties exist. The diversity of HTA activities across the member states has resulted in differences in the sustainability assessment of health technologies which have contributed to inequalities in the access and reimbursement decisions (Pereira et al., 2021). For example, a review of the HTA recommendations for orphan drugs in four European countries found considerable variation in both the decision-making criteria and the outcomes, which indicate the lack of a more holistic mechanism (Picavet et al., 2013).

In the EU framework on HTA, stakeholder engagement is very important, as it guarantees that all relevant parties, such as patients, health-care providers, industry, and government, are sufficiently interested in the evaluation. This engagement across multiple sectors is key in evaluating health technologies and informing policy decisions (Van Haesendonck, 2023; Desmet, 2024).

According to the European Access Academy, stakeholders have provided relevant feedback on their functions and expectations in the context of the emerging EU HTA system. While focusing on the developing debates and conventions, some surveys have shown that stakeholders should be actively engaged during the beginning stages of the evaluation process to promote confidence and openness (Van Haesendonck, 2023; Desmet, 2024). The creation of civil society dialogues is also encouraged to make sure that all stakeholders have a voice during the HTA process (Desmet, 2024). There is a likelihood of conflicts and contradictions (Pereira et al., 2021). I have some doubts about the practicalities of it.

Furthermore, the prerequisites established in the EU HTA system might be too much for some less-developed countries, and there will be a necessity for support from the more sophisticated HTA countries (Ruether and Petelos, 2022).

On the contrary, the EU regulation on HTA has great potential to increase the efficiency of HTA. The promotion of joint clinical assessments (JCA) and the sharing of methodologies and data accomplishes the regulation's objectives of decreasing the work duplication and decisional level bottlenecks (Ruether and Petelos, 2022).

With advances in new therapies, the EU HTA framework needs to be adjusted in a manner that ensures that assessments are current, relevant, and robust (Iandolo et al., 2018). Also, the incorporation of RWE and PROs into the HTA is now a necessity. These factors can provide great insight into the day-to-day operations of health technologies and their effectiveness in clinical settings (Iandolo et al., 2018). For a cost-utility analysis in the context of HTA, see, for example, Kennedy-Martin et al. (2020).

References

Berntgen, M., et al. (2014). Improving the contribution of regulatory assessment reports to health technology assessments–A collaboration between the European medicines agency and the European network for health technology assessment. *Value in Health, 17*(5), 634–641.

Cook, C. E. (2008). Clinimetrics corner: The minimal clinically important change score (MCID): A necessary pretense. *Journal of Manual & Manipulative Therapy, 16*(4), E82–E83. https://doi.org/10.1179/jmt.2008.16.4.82E

De Vet, H. C. W., Terwee, C. B., & Mokkink, L. B. (2023). Minimal important difference. In F. Maggino (Ed.), *Encyclopedia of quality of life and Well-being research*. Springer. https://doi.org/10.1007/978-3-031-17299-1_1813

Desmet, T., Julian, E., Van Dyck, W., Huys, I., Simoens, S., Giuliani, R., Toumi, M., Dierks, C., Dierks, J., Cardone, A., Houÿez, F., Pavlovic, M., Berntgen, M., Mol, P., Schiel, A., Goettsch, W., Gianfrate, F., Capri, S., Ryan, J., Ducournau, P., et al. (2024). An inclusive civil society dialogue for successful implementation of the EU HTA regulation: Call to action to ensure appropriate involvement of stakeholders and collaborators. *Journal of Market Access & Health Policy, 12*(1), 21–34. https://doi.org/10.3390/jmahp12010004

Edinoff, A. N., Nix, C. A., Hollier, J., Sagrera, C. E., Delacroix, B. M., Abubakar, T., Cornett, E. M., Kaye, A. M., & Kaye, A. D. (2021). Benzodiazepines: Uses, dangers, and clinical considerations. *Neurology International, 13*(4), 594–607. https://doi.org/10.3390/neurolint13040059

Engel, L., Beaton, D. E., & Touma, Z. (2018). Minimal clinically important difference: A review of outcome measure score interpretation. *Rheumatic Disease Clinics of North America, 44*(2), 177–188. https://doi.org/10.1016/j.rdc.2018.01.011

European Medicines Agency. (2021). *Spinraza: conditional marketing authorisation.* Retrieved from EMA website.

FDA. (2009). *Guidance for industry: Patient-reported outcome measures: Use in medical product development to support labeling claims.* Food and Drug Administration.

Fekete, I., Hall, S., Bohn, J., Paradis, A., Schneller, S., Gleißner, E., Coratti, G., Bovis, F., Nascimento, A., Povedano, M., Sormani, M. P., Vázquez, J., Mercuri, E., Kirschner, J., Patel, S. N., & Volmer, T. (2020). Statistical solutions for analytic challenges in registries for HTA-purposes. *Value in Health, 23*(Supplement 2), S692.

Fontrier, A. M., Visintin, E., & Kanavos, P. (2022). Similarities and differences in health technology assessment systems and implications for coverage decisions: Evidence from 32 countries. *Pharmaco Economics – Open, 6*(3), 315–328. https://doi.org/10.1007/s41669-021-00311-5

Franceschini, M., Boffa, A., Pignotti, E., Andriolo, L., Zaffagnini, S., & Filardo, G. (2023). The minimal clinically important difference changes greatly based on the different calculation methods. *The American Journal of Sports Medicine, 51*(4), 1067–1073. https://doi.org/10.1177/03635465231152484

Graili, P., Guertin, J. R., Chan, K. K. W., & Tadrous, M. (2023). Integration of real-world evidence from different data sources in health technology assessment. *Journal of Pharmacy & Pharmaceutical Sciences: A Publication of the Canadian Society for Pharmaceutical Sciences, Societe Canadienne Des Sciences Pharmaceutiques, 26*, 11460. https://doi.org/10.3389/jpps.2023.11460

Harrison, C., Kiladjian, J., Al-Ali, H. K., Gisslinger, H., Waltzman, R. J., Stalbovskaya, V., et al. (2012). Jak inhibition with ruxolitinib versus best available therapy for myelofibrosis. *New England Journal of Medicine, 366*(9), 787–798. https://doi.org/10.1056/nejmoa1110556

Iandolo, F., Vito, P., Fulco, I., & Loia, F. (2018). From health technology assessment to health technology sustainability. *Sustainability, 10*(12), 4748. https://doi.org/10.3390/su10124748

IQWiG. (2023). *General methods version 7.0*, https://www.iqwig.de/methoden/general-methods_version-7-0.pdf

Julian, E., Pavlovic, M., Sola-Morales, O., Gianfrate, F., Toumi, M., Bucher, H. C., Dierks, C., Greiner, W., Mol, P., Bergmann, J. F., Salmonson, T., Hebborn, A., Grande, M., Cardone, A., & Ruof, J. (2022a). Shaping a research agenda to ensure a successful European health technology assessment: Insights generated during the inaugural convention of the European access academy. *Health Economics Review, 12*(1), 54. https://doi.org/10.1186/s13561-022-00402-x

Julian, E., Gianfrate, F., Sola-Morales, O., et al. (2022b). How can a joint European health technology assessment provide an 'additional benefit' over the current standard of national assessments? *Health Economics Review, 12*, 30. https://doi.org/10.1186/s13561-022-00379-7

Kennedy-Martin, M., Slaap, B., Herdman, M., et al. (2020). Which multi-attribute utility instruments are recommended for use in cost-utility analysis? A review of national health technology assessment (HTA) guidelines. *The European Journal of Health Economics, 21*, 1245–1257. https://doi.org/10.1007/s10198-020-01195-8

Larose, H., Lee, M., Grueger, J., et al. (2023). Opportunities to improve the adoption of health-related quality of life evidence as part of the French health technology assessment process. *Health Res Policy Syst, 21*, 137. https://doi.org/10.1186/s12961-023-01081-8

Mercuri, E., Finkel, R. S., Muntoni, F., Wirth, B., Montes, J., Main, M., et al. (2018). Diagnosis and management of spinal muscular atrophy: Part 2: Treatment, care, and therapeutic strategies. *The Lancet Neurology, 17*(9), 735–747. https://doi.org/10.1016/S1474-4422(18)30293-7

Nielsen, C., Merrell, D., Reichenbach, R., Mayolo, P., Qubain, L., & Hustedt, J. W. (2023). An evaluation of patient-reported outcome measures and minimal clinically important difference usage in hand surgery. *Plastic and Reconstructive Surgery. Global Open, 11*(12), e5490. https://doi.org/10.1097/GOX.0000000000005490

Pereira, et al. (2021). HTA reshaping: Rethinking the health technology assessment framework in Portugal. *Portuguese Journal of Public Health, 39*(1), 36–47. https://doi.org/10.1159/000516501

Picavet, E., Cassiman, D., Hollak, C. E., et al. (2013). Clinical evidence for orphan medicinal products-a cause for concern? *Orphanet Journal of Rare Diseases, 8*, 164. https://doi.org/10.1186/1750-1172-8-164

Ruether, A., & Petelos, E. (2022, October). The development of HTA collaboration at European level and the challenge of the EU regulation. *European Journal of Public Health, 32*(Supplement_3), ckac129.086. https://doi.org/10.1093/eurpub/ckac129.086

Ruether, A., Imaz-Iglesia, I., Bélorgey, C., Lo Scalzo, A., Garrett, Z., & Guardian, M. (2022). European collaboration on health technology assessment: Looking backward and forward. *International Journal of Technology Assessment in Health Care, 38*(1), e34. https://doi.org/10.1017/S026646232200006X

Schlichting, M., Hennig, M., Rudell, K., McLeod, L., Bennett, B., Shaw, J., Doward, L., Molsen-David, E., & Chassany, O. (2022). Is IQWiG's 15% threshold universally applicable in assessing the clinical relevance of patient-reported outcomes changes? An ISPOR special interest group report. *Value in Health: The Journal of the International Society for Pharmacoeconomics and Outcomes Research, 25*(9), 1463–1468. https://doi.org/10.1016/j.jval.2022.07.010

Smart Step Consulting. (2023). *The number "15" in the German HTA system–the "fixed threshold" as a fixed solution?* Retrieved from https://www.smartstep-consulting.de/en/the-number-15-in-the-german-hta-system-the-fixed-threshold-as-a-fixed-solution/

The Federal Joint Committee (G-BA)

Urbina, I., Adams, R., Fernandez, J., Willemsen, A., Hedberg, N., & Rüther, A. (2024). Advancing cooperation in health technology assessment in Europe: Insights from the EUnetHTA 21 project amidst the evolving legal landscape of European HTA. *International Journal of Technology Assessment in Health Care., 40*(1), e75. https://doi.org/10.1017/S0266462324004689

Van Haesendonck, L., Ruof, J., Desmet, T., Van Dyck, W., Simoens, S., Huys, I., Giuliani, R., Toumi, M., Dierks, C., Dierks, J., Cardone, A., Houÿez, F., Pavlovic, M., Berntgen, M., Mol, P. G. M., Schiel, A., Goettsch, W., Gianfrate, F., Capri, S., Ryan, J., et al. (2023). The role of stakeholder involvement in the evolving EU HTA process: *Insights generated through the European access academy's multi-stakeholder pre-convention questionnaire. Journal of Market Access & Health Policy, 11*(1), 2217543. https://doi.org/10.1080/20016689.2023.2217543

Wood, S.N. (2017) Generalized Additive Models: An Introduction with R. Second Edition, CRC Press, Boca Raton. https://doi.org/10.1201/9781315370279

Chapter 7
The Role of Minimal Important Difference (MID) in Drug Development Process

Abstract The process of drug discovery and early development occurs in a staged manner. It starts with considerable safety screening and potential efficacy assessment of new compounds. A blend of in vitro and in vivo models is used to assess the pharmacological action and toxicity of a drug, as well as its pharmacokinetics. Risk assessment in this phase utilizes several quantitative thresholds for evaluating toxicity in a tiered approach, such as the no observed adverse effect level (NOAEL), median lethal dose (LD_{50}), half maximal inhibitory concentration (IC_{50}), and effective concentration 50 (EC_{50}). These thresholds are pivotal to inform important safety decision requirements for further testing in first-in-human (FIH) clinical trials.

FIH trial is the first test in humans, marking a key checkpoint in the drug development process. These studies focus on safety, tolerability, pharmacokinetics, and pharmacodynamics, evaluating all these in dose-escalation frameworks of single ascending dose (SAD) and multiple ascending dose (MAD). The preclinical studies NOAEL, human equivalent dose (HED) and maximum recommended starting dose (MRSD) derivations provide baseline estimates for FIH dosing levels, which depend on a safe and effective dose to commence. These estimates come from allometric scaling and application of safety factors.

In early-phase development, the concepts of minimal clinically important difference (MCID) or minimal important difference (MID) usually come into play in much later-stage trials due to their focus on patient-reported outcomes (PROs) and effectiveness measures. Nonetheless, incorporating MCID reasoning into first-in-human (FIH) studies can enhance dose selection and pharmacodynamic (PD) predictive models.

When PD models detect validated clinical milestones early, forecasting which drug candidates to de-risk becomes more precise. This is crucial in oncology, where surrogate markers like tumor size reduction or progression-free survival (PFS) are often linked to established MCID thresholds.

Organ-on-a-chip and 3D microtissues, such as human liver models, have advanced the in vitro to in vivo translation of toxicity thresholds into clinical relevance. These models help define important clinical thresholds for metrics such as drug-induced liver injury (DILI), which is critical in the safety evaluation of new drugs. While referred to as ROC curve analysis, predictive validity and optimization

I. Fekete, *Transforming Data into Informed Decisions across Clinical and Non-Clinical Domains*, https://doi.org/10.1007/978-3-032-02595-1_7

of sensitivity/specificity, reasoning similar to MCID estimation is employed, though often not labeled as such.

In summary, early drug discovery and development steadily appreciates the necessity of defining MCIDs, as they build on concepts that surpass simple statistically significant thresholds. Implementing patient-centric principles such as MCID during preclinical and FIH stages represents a dramatic change toward more responsible and streamlined drug development frameworks.

Keywords MID in drug development · Preclinical trials · Clinical trials · MCID threshold mappings · Drug efficacy · MID in pharmaceutical research · FIH study

The Role of MID in Drug Development and During Preclinical and Clinical Trials

In drug discovery, cell-based (in vitro) and animal-based (in vivo) studies are conducted. Their objective is to assess the safety, pharmacokinetics, and pharmacodynamics of potential compounds. The preclinical step is characterized by threshold-based metrics quantitatively estimating the bounds of safety and efficacy of the research medicine under investigation.

Some of the numerous thresholds that are examined include: (i) no observed adverse effect level (NOAEL), (ii) median lethal dose (LD_{50}), (iii) half maximal inhibitory concentration (IC_{50}), and (iv) effective concentration 50 (EC_{50}). For an overview of these measures, Walum (1998) is a good source. In this chapter, I will discuss preclinical phases of drug development, particularly focusing on these thresholds.

To exemplify, median lethal dose (LD_{50}) is the dose that causes 50% of the animals to die in a given time due to acute toxicity (Walum, 1998). The preclinical phase is designed to foresee and simulate the response of a drug in animals prior to any actual clinical interactions. This prior phase has two major objectives: first, it asserts proof of concept with respect to the pharmacological activity of the drug, and second it estimates its safety through relevant toxicological studies.

Such studies are prerequisites for the new drug application (NDA) or investigational new drug (IND) filings with the FDA or EMA. This phase consists of multiple evaluations, acute, sub-chronic, and chronic toxicity, genotoxic and carcinogenic toxicity, reproductive and developmental toxicity, ADME (absorption, distribution, metabolism, and excretion).

At each of these stages, quantifiable decision-making thresholds are established. Unfortunately, their threshold setting is different from MID/MCID. Hence, bridging these two clinical phases is difficult. However, these thresholds are of great importance in ascertaining dose levels for FIH trials (First-in-Human trials) and as clinical risk mitigation strategies to ascertain that no undue harm is administered to potential participants of the trials. This is where MCID could come in handy.

Traditionally, drug development and MCID considerations are linked to the later phases of development, particularly in Phase II/III trials. However, there is a strong case to be made for including MCID/MID considerations even during FIH studies to improve the entire drug development continuum. The primary objectives of FIH studies are to: evaluate safety, tolerability, pharmacokinetics (PK), and pharmacodynamics (PD), determine and characterize the dose–response relationship for the drug, and provide information for dose escalation and the design of subsequent Phase II studies.

The question arises as to how to integrate MCID/MID with FIH design. Clinically important changes do not apply in the context of FIH trials; however, predicting them beforehand using anticipated MCID values helps in creating useful PK/PD models: a dose that is biologically active will not be worthwhile if the expected efficacy goals (such as symptom reduction or decrease in relevant biomarkers) do not meet the projected dose that purportedly would lead to MCID. This supports "go/no-go" decisions being made early, which improves efficiency.

Second, PD marker or surrogate endpoints are often captured in FIH trials. If these PD markers can be tethered to clinical endpoints anchored to an MCID based on historical data or RWE, they enable the validation of early signals. Third, for example, clinically, a 10 mmHg decrease in blood pressure may serve as the MCID in terms of mitigating cardiovascular risk. In a situation where the pharmacodynamic effect in FIH demonstrates only a 3–5 mmHg reduction even at the dose of MD, it would not be warranted to pursue further investment. Moreover, dose selection based on the target product profile (TPP) typically outlines certain parameters, including the efficacy profile. As such, MCID is inherently linked to the TPP. FIH studies should evaluate whether any of the early dosing strategies fall within the treatment profile defined by MCID.

For example, in oncology, tumor size reduction or progression-free survival (PFS) is frequently utilized as a surrogate endpoint in early-phase studies. Moreover, MCIDs have been estimated for PFS (e.g., 2–3 months for certain malignancies) and are also informed by PROs such as pain and fatigue. In FIH oncology trials, PD markers such as tumor reduction at various dose levels can be plotted against these MCID threshold mappings: if the maximum tolerated dose (MTD) results in a tumor response that is less than the MCID, this may indicate lack of efficacy.

Considering MCID/MID in FIH trial design is an evolving paradigm in the translational drug development scaffolding. In spite of the fact that FIH studies are not intended to capture clinical efficacy, they serve as the initial, practical evaluation of a molecule's potential to achieve clinically meaningful effects. If early clinical pharmacologic signals are rationally timed to later-stage milestones informed by MCID concerns, then resource allocation, ethically plausible justifications, and attrition rates in drug development and in the entire portfolio of controlled trials could be improved.

The processes of identifying and designing new drugs is called drug discovery, and it requires new therapeutics to have strategies and plans that guarantee the evaluations conducted are impactful and practical, and, last but not least, *safe*. While exploring the parameters involved in drug discovery, MID/MCID is formed at a later stage, which seeks to set meaningful thresholds that entail substantial alterations enhancing defined value metrics when critically examining health outcomes with regard to patient care and toxicity.

FIH trials are pivotal for any investigational medicine, as they signal the change from foundational work to real-world testing on patients. It is the first form of clinical testing on human participants, where a new drug (or medicinal product), biological or otherwise, is put to FIH. Prior to FIH trials, all testing is done in vitro or in vivo, as mentioned earlier, in a laboratory setting or on animals. Hence, compared to other testing, FIH trials are the first point of interaction with humans and allow for the assessment of how the compound interacts with human physiology and biology.

Assessing the impact of the drug on humans is the main goal of any FIH trial, alongside measuring *safety*, *tolerability*, and *effectiveness*, thus making it possible to gauge potential risks. FIH studies are set up to attempt and assess every possible adverse reaction that may occur to establish safety in risk-free hazard-free environments.

Determining dose levels is an integral component of an FIH study design, with goals including determining the maximum tolerated dose (MTD), minimally effective pharmacological dose (if any), and appropriate starting dose. Usually, studies use a dose-escalation strategy that consists of single ascending dose (SAD) and multiple ascending dose (MAD) arms. These dose levels are implemented to provide controlled clinical exposure in a stepwise manner. Generally, FIH trials have small sample sizes and usually range from 20 to 100 participants, comprising primarily of healthy volunteers or, in some cases, patients with the condition of interest.

Participants undergo advanced physiological surveillance, including ECGs, laboratory tests for blood chemistry, and other vital sign monitoring, alongside continuous observation, ensuring rapid response to any emerging safety concerns.

FIH trials are governed within an explicit ethical and regulatory boundary. They operate within the framework of good clinical practice (GCP) and undergo approval by regulatory bodies, which may include the FDA, the EMA, or the Drug Controller General of India (DCGI).

The potential human participants must not be exposed to any form of the investigational product, which must undergo toxicology assessment in at least two different classes of animals: a rodent and a non-rodent. Some critical thresholds must be set, for example, those mentioned earlier: NOAEL, LD_{50}, IC_{50}, and the EC_{50}.

Ranging from the safe starting dose to the margin of safety, these thresholds determine the quantitative estimate of the risk of human exposure. These thresholds aid in the rudimentary stages of clinical trials and provide a framework to quantify dose estimation in human testing.

The UK TGN1412 FIH trial tragedy in 2006 serves as a benchmark case in which healthy volunteers encountered life-threatening cytokine storms for 5 days into the trials and were thereafter put on life support. Exemplifying an important case relating to the risks associated with human clinical trials is the TGN1412 study.

As mentioned earlier, FIH are the primary trials framed within other established guidelines that serve as checkpoints to prevent complete failure. They help achieve needed benchmarks along parallel rails at hand, which can keep the chances of experimental failure in check.

It goes without saying that FIH marks pivotal milestones on the continuum of drug development. They are points of access for the remaining clinical phases because successfully passing through phase FIH is a mandatory requirement to proceed into phase II where the challenge is centered around efficacy and safety.

To sum up, ICH (2009) comes up with the following thresholds as illustrated in Table 7.1.

Moreover, there is LOD which is defined as the lowest quantitative level that can be measured with a CI above a certain threshold. It is universally accepted that the LOD should indicate the minimum concentration or quantity of the analyte present. Phase I clinical trials seek to find the maximum tolerated dose (MTD) of a drug. This is critical for its further development in later phases (Le Tourneau et al., 2009).

In my opinion, the BMD should be combined with the MCID in that it should be tailored to the needs of the patient. In its current form, the defined range is not based on MCID.

Speaking of toxicity, BMD is a promising concept in preclinical testing too, where new modeling approaches are constructed. Let me exemplify it with the assessment of drug-induced hepatotoxicity/liver-injury (DILI) via liver microtissues (Proctor et al., 2017). I would like to show the use of thresholds in the field of drug discovery by employing this example. The reader should ponder MID in this context.

If adopted, appropriate in vitro biological models may enable scientists to simulate the responses of humans to medications, which is crucial to minimizing the high attrition rates witnessed in clinical trials.

Table 7.1 The main thresholds based on ICH (2009). The table sketches a side-by-side look at the dose–response benchmarks that drive preclinical safety profile and the earliest stages of human testing, all viewed through the lens of the ICH M3(R2). In each row a standard toxicological or pharmacological term is named, its practical sense is laid out, and its precise standing under ICH M3(R2)-the guideline that dictates when non-clinical studies must wrap-up before a FIH trial-is noted

Term	Definition	Relevance in ICH M3(R2)
NOAEL	Highest dose with no adverse effect	Primary threshold for starting dose estimation
LOAEL	Lowest dose with an observed adverse effect	Used when NOAEL is not available
MTD	Highest dose that does not cause unacceptable side effects	Important in oncology drug development
PAD	Lowest dose that produces desired pharmacological effect	Alternative to NOAEL in mechanism-based toxicity
BMD	Dose producing a defined change in response (e.g., 14%)	Statistically modeled, increasingly preferred over NOAEL

Human micro-liver tissue possesses predictive capabilities for DILI, an important consideration in the *discovery* and *safety evaluation* of drugs in the industry. The research by Proctor et al. (2017) elaborates on the importance of using 3D liver models and how they improve accuracy in predicting hepatotoxicity, a key factor in developing new medicines. I would like to show you the results from Proctor et al. (2017), performed on 110 marketed drugs. The figures are copyrighted by Proctor et al. (2017).

The question arises as to how Proctor et al. (2017) determine thresholds, and how their method(s) relate to MIDs. While not providing specific biochemical benchmarks for DILI in their study, Proctor et al. (2017) highlight some broadly accepted thresholds. Within the literature, there are baseline benchmarks that are considered guidelines for harm evaluation, and these usually require evaluation by an external authority.

As noted earlier, Proctor et al. (2017) studied the predictive validity of 3D human liver microtissues regarding DILI—a fundamental question in both preclinical toxicology and pharmaceutical development. These aspects aid in the advancement of the in vitro–in vivo translation frameworks, which use laboratory models for clinical systems, determining whether hepatotoxicity observed in the system will result in clinically meaningful toxicity in the population of interest.

What ties their work to MID is the thresholding step involved in their DILI signal evaluation: Proctor et al. (2017) had to set what levels of in vitro toxicity would be deemed "clinically concerning" and thus actionable. They were attempting to estimate predictive MIDs—the level at which in vitro changes (like ATP reduction or reduced spheroid viability) in a liver microtissue model could be interpreted as thresholds that reliably predict in vivo hepatotoxic outcomes.

While no explicit discussion of MID is provided, the reasoning behind setting bioassay cut-off levels to meaningful clinical outcomes closely resonates with the MCID framework. Moreover, the use of receiver operating characteristic (ROC) curves alongside sensitivity/specificity optimization in Proctor et al.'s (2017) study serves as a proxy for MID estimation.

As an example, the authors selecting the best cut-off for predicting DILI asked in essence: what is the lowest level of in vitro error that distinguishes between toxic and nontoxic substances in a clinically pertinent way? In my opinion, this is similar to how MID is employed in PROMs or functional status instruments to clinically anchored research.

As noted by Proctor et al. (2017), these thresholds are not dependent on statistical validity alone, but rather on whether they improve decisions, decrease unnecessary harm, and support overall clinical judgment of importance.

In most cases, the established clinical and DILI benchmarks set by guiding authorities are the ones used. These are based on serum levels of aminotransferases such as ALT of and ALP. Thus, there is an upper limit of normal (ULN) or cut-off, which is defined as the limit not exceeded (ULN), and any level above that is positive.

In the studied pharmacological sets, Proctor et al. (2017) put forth recommendations for standardized thresholds that serve as cut-off values. For example, based on clinical evidence, *fivefold elevation of ALT* over the upper limit qualifies as a standard measure for severe hepatocyte damage, while for *ALP*, a *twofold increase* over the upper limit is considered a marker of liver damage associated with cholestatic process and is widely accepted.

In addition to these, liver toxicity associated with heightened *ALT exceeding threefold rise* above three the ULN (upper limit of normal) and elevated total bilirubin mark are frequently cited references. These differentiators are fundamental to assessing clinical management practices and form the underlying guiding principles for differentiating liver damage from the multitude of liver injury. Figure 7.1 illustrates that DILI severity grades. On the *x*-axis, the different drugs are displayed.

In the scope of DILI, setting criteria is vital in finding and categorizing cases of liver injury stemming from the use of medicinal drugs. A benchmark is set with ALT (alanine aminotransferase) being equal to or exceeding five times excess the upper limit of normal (ULN).

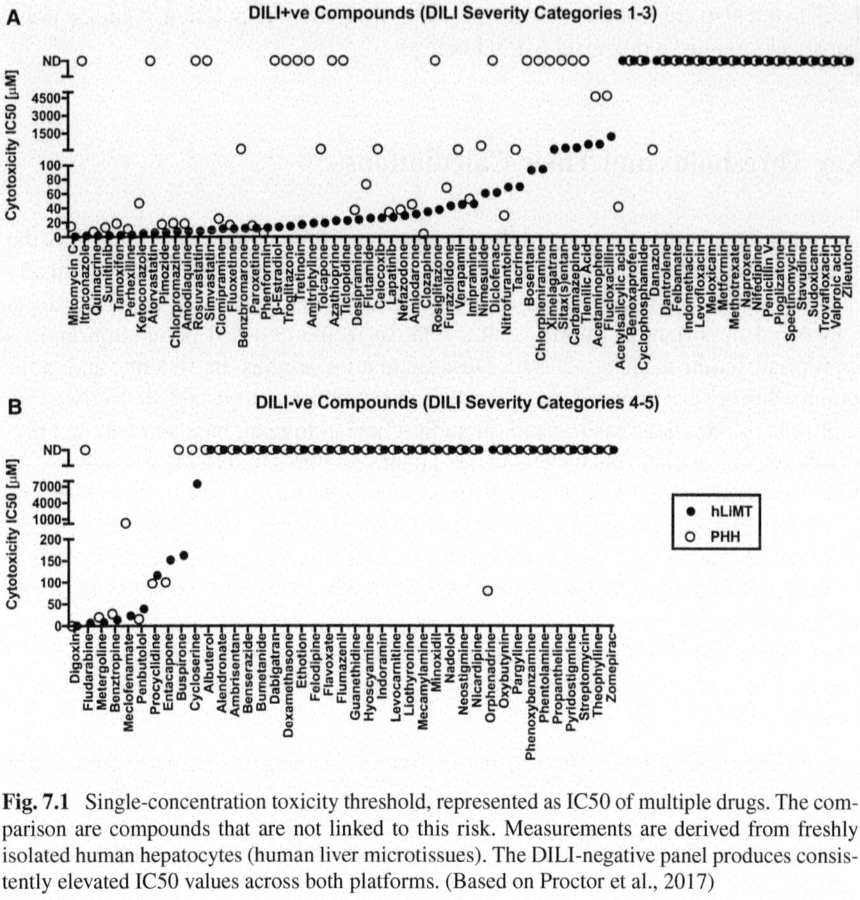

Fig. 7.1 Single-concentration toxicity threshold, represented as IC50 of multiple drugs. The comparison are compounds that are not linked to this risk. Measurements are derived from freshly isolated human hepatocytes (human liver microtissues). The DILI-negative panel produces consistently elevated IC50 values across both platforms. (Based on Proctor et al., 2017)

Moreover, the specific thresholds are affected by the type of patient and their specific liver enzyme levels, once again highlighting interindividual variability. For example, some criteria may shift because of genetic factors or comorbidities that are likely to cause liver injury, leading to the debates regarding individualized definitions and approaches for such conditions.

As already discussed, although most of the literature has been found to follow set thresholds, the intricacies and divergences of DILI require that a reasoned approach consider *the individual patient*, the *drug in use*, and the possible pathophysiological relationships existing borders and intersections between these components. A greater partnership in the form of a joined effort without pre-discussed boundaries, among researchers, clinicians, and legislators is needed to establish these limits in order to manage and accurately diagnose cases of DILI effectively.

Particularly in drug development, including oncology, the importance of setting appropriate limits is critically established. Pharmacogenetic studies confirm that differences in one's genetic profile can determine how patients respond to some therapies. I regard this as another argument against a universal MCID. This explains the need for more precise and patient-oriented MID values that incorporate not only the data but also clinical knowledge regarding the patient population. Again, a point that speaks against a universal MID, I believe.

Key Thresholds and Their Calculations

The most frequently encountered thresholds in preclinical pieces of research is the NOAEL, which refers to the highest exposure level at which there is no statistically or biologically significant increases in the frequency or severity of adverse effects considered important to public health, relative to the exposed population and its appropriate control (ICH, 2009). Dose-escalation studies in rodents and non-rodents, lasting between 14 and 90 days, are conducted to establish NOAEL. A multitude of parameters is tracked, including hematological, biochemical, and histopathological markers, as well as various doses administered to the animals.

The calculation of NOAEL is an estimation procedure that requires comparisons of treated and control groups using traditional statistical hypothesis tests (e.g., ANOVA, Dunnett's test) and verification through toxicological expert judgement. As the result of such calculations, the NOAEL is often presented with LOAEL estimate and then the range of toxicity can be provided. NOAEL/human exposure is also used without dividing and regarded as safety risk estimation using simply the margin of exposure: lower ratios suggest an elevated toxicological concern. (Gad, 2016).

The next threshold is the median lethal dose (LD_{50}). LD_{50} refers to the dose required to kill 50% of a test population and has historically served as an index of acute toxicity. The concern now is ethical, and alternative methods need to be implemented, but it remains a reference point for certain chemical risk assessments.

The computation of LD_{50} is usually derived from a dose–response curve using probit or logit modeling, followed by interpolation to estimate the dose that results in 50% mortality rate. The thresholds of LD_{50} assist in determining the intents for description, classification, and management of the compound within the confines of regulatory policies, including REACH (Registration Evaluation Authorization and Restriction of Chemicals).

Third, the half maximal inhibitory concentration (IC_{50}). IC_{50} is defined as the concentration required to inhibit a certain biological activity (i.e., enzyme activity, cell proliferation) to 50% of its baseline level. This value is significant in pharmacodynamics and in preliminary screens determining the effectiveness of a drug.

IC_{50} estimations are achieved through nonlinear regression models, typically using four-parameter logistic regression of data gathered from constructing dose-response curves. Based on the definition of IC_{50}, dose-concentration biological activity response curves must be constructed, with *IC_{50} being* the 50% inhibitory concentration on the curve.

Fourth, the effective concentration 50 (EC_{50}) indicates the concentration of a drug needed to attain half of its maximal effect. It serves as a benchmark for compounds that act as agonists or stimulants, not inhibitors. While EC_{50} shares similarity with IC_{50} in the methodology of calculation, the former centers on the efficacy quotient. The threshold EC_{50} is calculated using the same type of sigmoid Emax model as in the case of IC_{50}.

One of the most important preclinical thresholds is their use for calculating starting doses in human trials. The FDA suggests using NOAEL (converted by allometric scaling) and applying interspecies safety factors (often between 10 and 100), which accounts for variability between species. This results in a human equivalent dose (HED) and a maximum recommended starting dose (MRSD) (FDA, 2005). These methods exemplify the precise threshold estimation required to balance safety and efficacy in early-phase clinical trials.

Preclinical drug discovery is being advanced by new developments in organ-on-a-chip technologies and humanized mice. These platforms permit more accurate thresholds that are more reflective of human biology.

Additionally, routine computation aids such as machine learning are being employed to estimate IC_{50} and NOAEL values based on chemical structures, streamlining primary screening and prioritization (Vamathevan et al., 2019). Vamathevan et al. (2019) propose that machine learning (ML) algorithms offer a solution for streamlined decision-making in drug discovery. Crucially, ML applications can be found at any phase of the drug development process. Examples include the validation of biological targets, discovery of prognostic biomarkers, and analysis of pathology in digitalized clinical trials (Vamathevan et al., 2019).

Nevertheless, some obstacles are yet to be addressed. Threshold values such as NOAEL are particularly prone to biases from study design limitations, differences between species, and cross-laboratory standardization issues. More importantly, the lack of patient-defined lower thresholds—similar to MCIDs—has been the focus of some debate in preclinical research. Enhancing credibility in translation requires a paradigm shift toward earlier integration of patient-centered concepts.

The ICH M3(R2) guideline defines the NOAEL as the highest tested dose or exposure level at which there is no increase, in some frequency or severity, to statistically or biologically important negative milestones when compared with an appropriate control group.

Critically, the concepts of NOAEL and lowest observed adverse effect level (LOAEL) differ. NOAEL conceptually differs from LOAEL in that it is the smallest dose at which some adverse outcome can be noted. In cases where a distinct NOAEL cannot be discerned—likely due to study design limitations or an uncertain toxicological profile—LOAEL can be utilized, albeit with a degree of caution.

Most recently, a Phase 1 FIH trial was finished. Preliminary results from the Phase I dose-escalation trial of mRNA-4359 showed potential for use in patients suffering from advanced solid tumors (King's College London, 2024). The focus of this investigational mRNA cancer immunotherapy is on lung, melanoma, and other solid tumors.

The results of the study were shared in Barcelona during a conference organized by the European Society of Medical Oncology. The primary investigator for the trial from King's College London and Guy's and St Thomas' NHS Foundation Trust presented these findings on September 14, 2025.

Returning to ICH M3(R2), it underscores the importance of the NOAEL in planning and conducting FIH clinical trials. NOAEL plays an important role as a baseline parameter for starting dose determination in FIH studies. The HED is usually obtained using allometric scaling with body surface area or with various pharmacokinetic modeling methods pertaining to the study drug (more on the study, see, EMA/CHMP/ICH/507008/2011 Committee for medicinal products for human use, CHMP).

In preclinical pharmacological studies, NOAEL has important applications in toxicological risk evaluation. Such a threshold serves as an important marker for potential safety limits. It also helps ensure that later clinical trials can be performed with adequate safety. Unlike other measures, though, the guideline equally considers the shortcomings of depending on NOAEL. In cases involving complex non-monotonic dose–response relationships or those requiring sophisticated risk quantification, alternative methods like benchmark dose (BMD) analysis or determining the pharmacologically active dose (PAD) can provide clearer insight.

Modern drug development begins with preclinical trials designed to gather important information about the safety and efficacy of new compounds. In this context, the estimation of quantitative thresholds such as NOAEL, LD_{50}, IC_{50}, and EC_{50} are crucial. Each threshold serves a specific purpose within the broader context of advancement into human testing.

In sum, the median effective dose is the dose at which 50% of the population shows some response or effect. The median toxic dose is the amount of a substance that is expected to cause a specified toxic effect in 50% of individuals. The median lethal dose is the amount of a substance needed to kill half of a specified animal population.

These thresholds are derived from statistical models, dose–response analyses, and expert evaluation. As the landscape of drug development evolves, ensuring that the refinement and individual tailoring of such thresholds accompanies ongoing change will be essential to the efficiency and ethical fabric of transnational medicine.

References

Gad, S. C. (2016). *Preclinical development handbook: Toxicology*. John Wiley & Sons.

International Conference on Harmonisation (ICH). (2009). *ICH harmonised tripartite guideline: Guidance on nonclinical safety studies for the conduct of human clinical trials and marketing authorization for pharmaceuticals M3(R2)*. Retrieved from https://database.ich.org/sites/default/files/M3_R2__Guideline.pdf

King's College London. (2024, September 13). *Experimental mRNA cancer vaccine shows potential for advanced stage cancer patients in phase 1 trial*. ScienceDaily. Retrieved May 6, 2025, from https://www.sciencedaily.com/releases/2024/09/240913145621.htm

Le Tourneau, C., Lee, J. J., & Siu, L. L. (2009). Dose escalation methods in phase I cancer clinical trials. *Journal of the National Cancer Institute, 101*(10), 708–720. https://doi.org/10.1093/jnci/djp079

Proctor, W. R., Foster, A. J., Vogt, J., Summers, C., Middleton, B., Pilling, M. A., Shienson, D., Kijanska, M., Ströbel, S., Kelm, J. M., Morgan, P., Messner, S., & Williams, D. (2017). Utility of spherical human liver microtissues for prediction of clinical drug-induced liver injury. *Archives of Toxicology, 91*(8), 2849–2863. https://doi.org/10.1007/s00204-017-2002-1

Vamathevan, J., Clark, D., Czodrowski, P., Dunham, I., Ferran, E., Lee, G., Li, B., Madabhushi, A., Shah, P., Spitzer, M., & Zhao, S. (2019). Applications of machine learning in drug discovery and development. *Nature Reviews. Drug Discovery, 18*(6), 463–477. https://doi.org/10.1038/s41573-019-0024-5

Walum, E. (1998). Acute oral toxicity. *Environmental Health Perspectives, 106*(suppl 2), 497–503. https://doi.org/10.1289/ehp.98106497

Chapter 8
MID Outside Medicine and Psychology

Abstract Identifying a change or difference as clinically significant goes hand in hand with the idea of minimal important difference (MID) in philology. MID can be extremely useful, as it helps identify the genuine change beyond statistical "random noise." It can be used to assist more efficient and fact-based approaches within the spheres of education, business, environmental studies, public policy, marketing, law, and other social services.

In environmental studies and climate science, assessing whether temperature or precipitation, MID can validate the true deviances. From a marketing standpoint, MID can validate whether MIDs appropriately justify another product investment. For businesses, perception-based metrics or changes, such as customer satisfaction or profit, can be gauged using MIDs, just to name a few areas.

The scope of MID encompasses the evaluation of innovations in teaching and the involvement of students in educational activities. If a new teaching approach is implemented, even if there are a statistically significant differences, these might not be meaningful. This becomes even more critical when cost–benefit considerations are made, or when interpreting student performance.

In public policy and policymaking, MID can assist in clarifying whether shifts in public perception and socioeconomic indicators like the unemployment rate or income inequality require policy intervention. As an example, a decrease in unemployment from 8.9% to 8.7% is likely to be considered statistically significant by many, but from a human value perspective (employment), it is utterly meaningless.

In user experience (UX) design and business, MID determines whether changes made to the interface of a website or to some of the features of the product can meaningfully improve user satisfaction. If stakeholders rely on *p*-values only, they will misinterpret trivial changes as impactful, resulting in poorly informed strategic decisions.

Restorative justice and psychological evaluation provide examples where MID is utterly important. MID lies within legal realms that have been hitherto underexplored in litigations. While it is uncommon to apply MID in legal contexts, it may assist in obtaining meaningful results in victim-offender mediation, plea bargains, or assessments of psychological harm. Psychologists can apply MID for quantifying the emotional or psychological damages, guiding the calculation of the

I. Fekete, *Transforming Data into Informed Decisions across Clinical and Non-Clinical Domains*, https://doi.org/10.1007/978-3-032-02595-1_8

appropriate sentence or compensation, thus integrating a rational, evidence-based framework into the process.

Finally, MID may also assess social programs, determining whether intended changes in the quality of life or the provision of services have been accepted as improvements by the intended users. In education, linguistics, health, user experience, and legal contexts, the common thread is that MID connects statistical data with the real world, thus reinforcing the credibility and usefulness of research and policy outcomes.

Keywords MID in philology · MID in nonclinical domains · MID in education · MID in linguistics · MID in business · MID in environmental studies · MID in public policy · MID in marketing · MID in law

The Need for MID in Nonclinical Domains

In the field of biological and climate science, for example, data analytics plays an important role in understanding and predicting the challenges of climate change. By analyzing vast amounts of environmental data, scientists can track climate patterns, predict future changes, or build time-series forecasting models. We need to understand what is considered a *random statistical fluctuation in precipitation or temperatures*, for example, in a time series, and what is regarded as *real shift or change* (sometimes referred to as *positive or negative "shock"* in the time-series literature). MID can come into the picture at this point. When I write "real," I mean a change or difference that exceeds the MID.

It should now be clear that in medical-psychological research, health care, and data-driven decision-making, MCID/MID has sometimes been an essential entity. However, the scope of this concept is far reaching and could be applied to a vast array of sectors, even from marketing through forestry to politics.

A marketing campaign could be tested using a simple research design: which of the following two products is better? If the difference is perceivable in terms of MID (one of the two products is preferred over the other one conspicuously), we can conclude that the difference is meaningful, and that one over the other product should be used.

Stakeholders, such as pharmaceutical companies, want to increase revenue. The same is true for insurance companies and hospitals too. This is where MCID/MID comes into the picture. All these actors want a surefire metric that guide their decisions. And these decisions should be optimal, for example, exceed the MCID/MID.

Hence, MCID/MID also finds its application in nonclinical studies in which analogous metric can help make informed decisions based on significant differences (between groups) or changes (between conditions). In the world of nonclinical studies, it can guide researchers, practitioners, and policymakers to understand the smallest change in an outcome that can be identified as important, noticeable, meaningful, substantial, or relevant, to name a few synonyms.

I have adopted the definition myself from the clinical literature. Nonclinical studies, such as those, for example, in the field of education sciences, politics, or social sciences, often deal with interventions that aim to improve QoL, or change of behavior.

Let us imagine a scenario in the *e-commerce industry*. Suppose a company is gathering data about their customers' shopping habits and preferences on a Likert scale. Suppose they want to identify what specific change in their service would make a noticeable difference to their customers. It would help the company to identify the MID to see if the change is really substantial or not. Is it worth investing money into the campaign, the product, etc.? When used properly, MID can also build trust between customers, the employees, and even the CEO.

The product "sells itself," so to speak, if the employees can argue that the present product or the effect has changed so much that customer satisfaction exceeded even the MID (which can be measured by a simple customer satisfaction questionnaire). But we have to presuppose that the customer is interested in MIDs and has belief in statistics (and, of course, some knowledge).

Similarly, in the *political arena*, employing MID is essential, in my mind. Politicians and policymakers have lots of data about *public sentiment, opinion polls, and socioeconomic indicators*. But the question is as to what specific change would result in a practical and perceivable change in the lives of the people. Again, this is the MID that politicians (i.e., their experts) need to determine, so that they can formulate better policies.

As an illustration, information regarding income, unemployment, and education can be utilized to formulate strategies toward reducing the income gap and increasing educational access in India, for instance. MIDs, which are insensitive to large samples, could evaluate the results of these policies.

In the case of education, MIDs would assist in designing a more impactful pedagogy and curriculum. For instance, educational institutions can analyze student achievement data to determine underperformance and design appropriate corrective measures.

Data can be used to evaluate the effectiveness of these interventions using MIDs, ensuring that the teaching strategies employed yield the desirable outcomes. Educators could use such data and MIDs to identify what specific change in their teaching approach would make a *noticeable difference in student engagement and learning outcomes* (we will go into more detail later).

Environmental conservation is another nonclinical domain where MID-based data-driven decision-making is a new area of research. For example, data on *deforestation rates*, *species population trends*, and *carbon emissions* provides valuable insights into the health of our ecosystems. This data is then used to inform conservation strategies, such as the designation of protected areas and the implementation of sustainable practices. MIDs could be beneficial in such matters.

In the domain of social services, data is being used to improve service delivery and client satisfaction, for instance. Data from multiple sources, such as the health care, the education sector, and social services, is used to "profile" vulnerable families with risk factors (no negative connotation intended in the use of the word "profile"). Social service agencies are employing data to understand the needs of the populations they serve better. This enables them to tailor their services to these needs of their clients.

But why do we need an analogous measure outside the fields of psychology and medicine, for example, in the business sector? Data can enhance the efficiency of decisions in industry, for example. By utilizing an MCID-like metric to streamline processes and eliminate inefficiencies, policymakers, or stakeholders can deliver care more effectively and at a lower cost. This can result, for example, in improved client satisfaction, or optimal reactions to changes in the industry.

Let us go back to the school scenario. Let us take a hypothetical study aimed at evaluating the effectiveness of a new teaching method on student performance compared to standard of teaching practice. A mere statistically significant increase in test scores may not truly reflect the impact unless it is *substantial enough* to be considered meaningful by the students or teachers as "anchors." Here, determining this threshold would involve understanding what degree of improvement would be considered significant or relevant by the stakeholders, teachers, or the parents. This would be a typical example of an anchor-based approach (Zhang et al., 2023).

In nonclinical studies related to, for example, environmental sciences, MID can help in assessing the impact of interventions on environmental quality. For instance, evaluating the effectiveness of a new waste management strategy in India, an MID-like metric can help determine the level of reduction in pollution levels that would be considered *meaningful by the local community*. Money, resources, and manpower invested: this understanding can help policymakers and funding agencies make informed decisions about which projects to support. If they just rely on traditional hypothesis testing (*p*-values), they might make sub-optimal decisions, let alone Type I error. The worst-case scenario is that if they rely only on *p*-values, they might invest wrongly. Consulting companies (especially those that mediate between pharmaceutical companies and national authorities) already use MCID when analyzing randomized controlled trials (RCTs). Finally, the process of determining an MID often involves consultation with stakeholders and sometimes, consensus building among experts (see, e.g., the Delphi method; Stone Fish & Busby, 2005).

Examples for the Application of MID in Other Disciplines

MID is crucial in understanding the impact and effectiveness of interventions or changes implemented in various nonclinical sectors, ranging from education and business through public policy to social sciences including, for example, in theoretical and applied linguistics.

Consider a hypothetical scenario in the education sector where a new teaching technique such as the so-called "flipped classroom" is introduced. A statistical analysis might reveal a significant difference in the students' performance scores or any other scores, for instance, satisfaction scores. Fahmi et al. (2024) employed the technique of flipped classrooms but did not compute MIDs. I tried to compute "the educational intervention effect threshold" (i.e., MID) for their work.

After the intervention, the median scores for the flipped classroom were typically around three points higher (Fahmi et al., 2024). A three- to five-point gain on educational evaluations such as the Test of English as a Foreign Language (TOEFL) reading sections can be regarded as significant with proper strategies and dedicated timeframes like 6-week programs. This is especially the case when it is confirmed that the changes are statistically significant alongside the changes described in student feedback ("anchor") concerning enhanced motivation and improved methods.

In clinical research, MCID typically ranges from 0.3 to 0.5 SD (Norman et al., 2003). Jaeschke et al. (1989) contend that "MCID was represented by mean change in score of approximately 0.5 per item, when responses were presented on a seven-point Likert scale."

One could determine if a three-point of four-point change is at least 0.3 SD if the SD of the score would have been given. Fahmi et al. (2024) do not report SDs, but in the reading sections of the TOEFL test, SDs are generally between five and ten points. Therefore, even a three-point difference with an SD of 10 and an MID of 0.3 SD can be deemed borderline MID.

However, we might find that this difference in another hypothetical study, although statistically significant, is not meaningful practically. The increase in scores might be too small to justify the resources invested in implementing the new strategy. In other words, it might not improve the students' understanding of the subject matter. The impact of the teaching intervention on students' overall educational experience and long-term learning outcomes must also be evaluated.

Moreover, the resources required for the new teaching method (e.g., training for teachers, new materials, time) must be weighed against the potential benefits. If the gains are marginal (though statistically significant), the investment might be better allocated elsewhere or not at all. We will talk about cost–benefit analysis later in another context.

In addition, there is another problem, namely, the extent to which the findings can be generalized to other contexts, subjects, student populations, countries, or cultures; this is also a critical consideration when evaluating MIDs in almost every scientific field. A teaching technique that shows statistically significant but minimal improvement in a highly controlled setting may not yield the same results in diverse educational environments.

Similarly, in the business world, MID can help organizations make informed decisions. Suppose a company introduces a new product feature that results in a slight increase in sales. While the difference might be statistically significant, it might not be practically meaningful when considering the costs associated with the development and marketing the new feature. By employing the MID concept, the company can evaluate if the change brought about by the new feature is meaningful enough to justify the investment and the costs.

MID can be seen as the smallest change in a business metric (such as sales or customer satisfaction) that is perceived as beneficial by stakeholders and justifies further investment in a product feature or strategy, considering the associated costs and benefits. In the business scenario described, implementing a new product feature that results in a *statistically significant* but *practically negligible* increase in sales may not warrant the associated development, investment, and marketing costs. By employing the MID concept, businesses can set an MID threshold that considers both statistical significance and practical importance (relevance, meaningfulness, etc.), ensuring that investments in new product features lead to meaningful improvements in business outcomes.

Most recently, a real-world example from the business field showing a change of statistical significance but not MID comes from the 2025 J.D. Power North America Airline Satisfaction Study (Power, 2025, May 7). (Specializing in consumer insights, J.D. Power is a prominent global data and analytics firm.)

The press release I am referring to is called "JetBlue Airways, Delta Air Lines and Southwest Airlines Rank Highest in Customer Satisfaction in Respective Segments" (Power, 2025, May 7). This study (Power, 2025, May 7) noted an improvement of six (sic) points in passenger satisfaction from the previous year (2024), which they claimed as an improvement on a 1000 (sic) scale. This change, although statistically significant, (the magnitude (extent) of change) suggests that there was a very small value added to the customer experience overall. Most probably, customers would not even perceive any difference (in terms of MID).

As cited by J.D. Power, in the 2025 study, the total passenger satisfaction score improved by six points relative to 2024. To some extent, this change was accomplished from a statistical standpoint that could be considered "significant;" however, the real-world importance for an improvement so small, especially on a scale of 1000 (sic), is arguable. An increase of six points equates to a 0.6% improvement (!), which is likely undetectable by/not perceivable to passengers and is likely below the MID threshold. Another issue that arises is why they used a 1000-point scale.

Moreover, their study is based on 10,224 (sic) passengers, which is a statistical concern given Type I error and p-inflation, as we have already talked about in previous chapters. Even after p-adjustment (correction for Type I error) or the employment of Bayesian statistics, the problem would still persist because it is not about the *p*-value. In my mind, given the 0.6% increase coupled with the 10,224 passengers is a typical example of random statistical fluctuation, and most importantly, Type I error. Instead of employing traditional statistics, J.D. Power could have employed Bayesian approaches, for example.

Further methodological and statistical and limitations undermine the rigor and interpretability of their results. I do not mean that United Airlines (801) and Air Canada (800) are indistinguishable (see Fig. 8.1), which should have been emphasized in the report. The study is based on self-reported information from 10,224 passengers—all of whom had flown recently. This sampling technique naturally results in a self-selection bias since it only covers those people who travel frequently and are likely to answer consumer polls. As a result, the results probably underrepresent rare travelers, who would have somewhat different opinions on customer satisfaction. Lack of openness about the sampling frame, randomizing techniques,

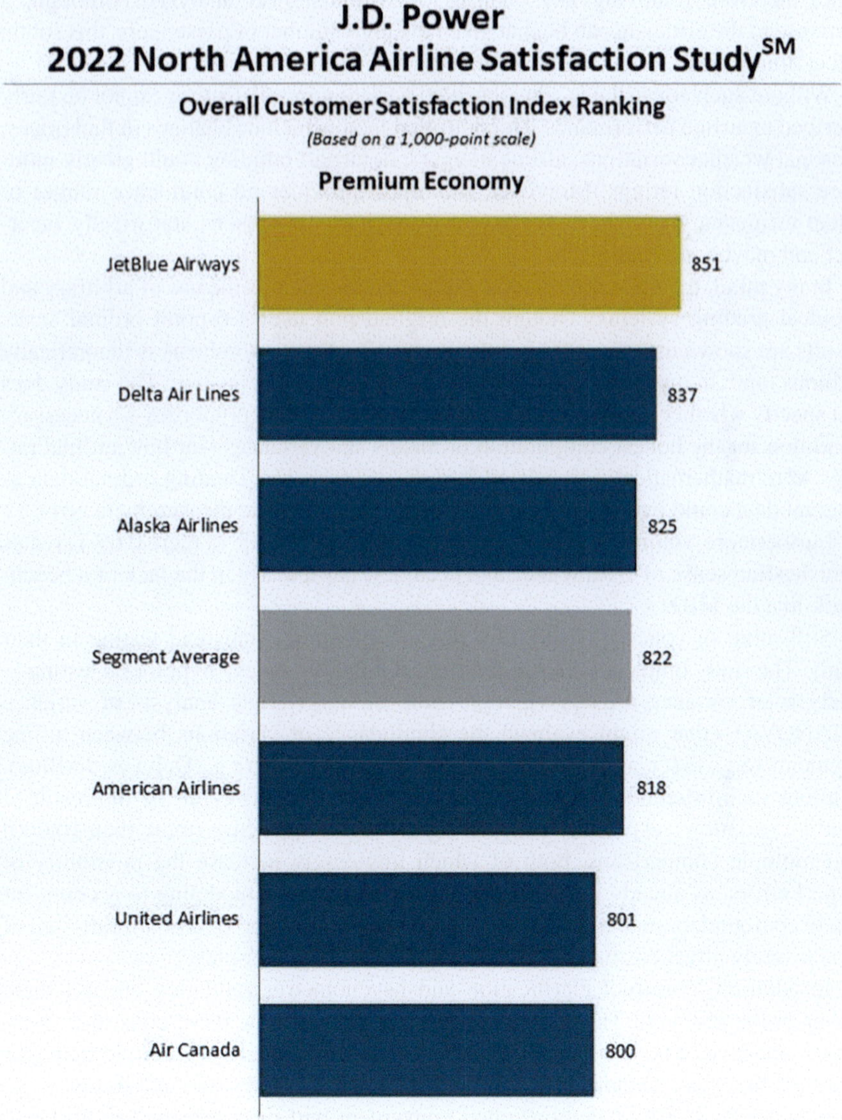

Fig. 8.1 The bar chart from J.D. Power's 2022 North America Airline Satisfaction Study demonstrates the overall customer satisfaction index ranking for airlines for premium economy service. The figure illustrates the 1000-point scale from 2022. (Copy righted by J.D. Power)

or demographic quotas calls into question how geographically and demographically accurate the sample is of the larger population of air travelers in North America.

Lack of longitudinal comparability and adjustment for confounders raises still another issue. The study carried out between 2024 and 2025 highlights a six-point rise in general, although it offers no proof that the same people were followed over time, therefore removing any chance for within-subject analysis. Although, a response to the criticism can be that given the high number of passengers, this subtle effect diminishes.

Without such consistency, any observed variation in satisfaction cannot be fairly ascribed to airline performance. Uncontrolled factors include changes in fuel prices, seasonal weather variations, airline mergers, or airport building could greatly influence satisfaction ratings. Moreover, the report provides no confidence ranges or effect estimates, therefore making year-over-year comparisons statistically incorrect and maybe misleading.

In my mind, the most serious methodological problem is the use of arbitrary and illogical grading systems. Though the original poll uses a 6-point ordinal scale, results are shown using a 1000-point measure. This metamorphosis is theoretically dubious (and, in my view, pointless) as well as statistically opaque. The study does not specify whether the new scale maintained interval-level properties—a necessary condition for the honest computation of means and variances—or how ordinal ratings were mathematically translated into the 1000 points. Treating ordinal data as interval data could overstate the effect magnitude and create measurement error.

Furthermore, stating a six-point gain on a 1000 point scale (a mere 0.6%) creates a misleading sense of significance and accuracy, particularly in the lack of a benchmark like the MID.

Still more of concern is the total lack of inferential statistical testing in their study. The study omits confidence intervals (CIs), SDs, or any hypothesis testing—analysis of variance (ANOVA), regression, or multivariate analysis of variance (MANOVA)—that might evaluate the significance of variations between airline segments (e.g., first class vs. economy) or between airlines (e.g., Delta vs. JetBlue). Without such inferential methods, any observed variations could be the result of random variation—especially considering the high sample size (more than 10,000) and multiple comparisons, both of which immensely increase the possibility of Type I errors, as already noted. Furthermore, multivariate modeling to account for likely confounders such as trip length, airport congestion, or kind of aircraft—all of which surely affect customer experience—is lacking in the study.

In addition, construct clarification and psychometric validation are not there either in the research. The report provides no data on how these constructs were operationalized, whether they were measured through single or multiple items, or whether they met standard benchmarks for dependability and reliability (e.g., Cronbach's $\alpha > .70$, or demonstrated convergent and discriminant validity), yet these constructs are central to the conclusions. This omission suggests that the scales might not fairly evaluate what they claim to assess, therefore raising the concern of instrumentation bias and design invalidity.

Temporal and contextual elements add to further confound the interpretation of the results. From March 2024 through March 2025, the poll was carried out most likely covering seasonal changes, during labor strikes, geopolitical disturbances, or travel restrictions linked to health. J.D. Power does not, however, provide any time breakdown, therefore hiding these factors. Such factors can greatly affect passenger satisfaction and should be included with time-series analysis or stratified reporting.

Moreover, a lost chance to gain greater insights from the dataset are the interaction effects—such as satisfaction varied by area, trip length, or airline hub status. Finally, its methodological defects—including nonrandom sampling, insufficient statistical testing, arbitrary score scaling, and lack of psychometric rigor—cause great vulnerability in its conclusions. Stakeholders may trust and act upon the results reached only by means of methodological robustness. Most importantly, I wanted to show how the absence of MID can "fool" people not trained in statistics.

We have already talked about public policy: MID can help in evaluating the effect of new policies or interventions. For instance, a policy aimed at reducing unemployment rates might show a statistically significant decrease in unemployment. However, an MID(-like) concept could help determine whether this decrease is meaningful in the real world. It might be that the decrease is too small to make a substantial difference in the lives of the unemployed, or the resources used to implement the policy could have been better utilized elsewhere.

A real-life example of a statistically significant but very small decrease in unemployment in India can be observed by examining the periodic reports and data released by the Centre for Monitoring Indian Economy (CMIE). The CMIE provides detailed and up-to-date labor force statistics, including unemployment rates in India. They write: "Unemployment rate in India, among persons aged 15 years and above, fell to 8.7% in December 2023 from 8.9% in the previous month. Although the unemployment rate eased in December, it remained at fairly high levels."[1] The terms "fell" and "eased" may convey a sense of substantial reduction in unemployment to a layman audience, which gives a distorted picture of reality. A more precise revision to mitigate this impression could be, if I may say so: "The unemployment rate in India experienced a *marginal decrease*, moving from 8.9% to 8.7%. This *minor shift*, while statistically significant, suggests only a *slight adjustment* in the overall employment landscape." (reformulation and emphasis mine).

According to this example of the CMIE, unemployment rate in India underwent a minor decrease during this time period. Nevertheless, this decrease is *statistically significant* but very small and *practically irrelevant*. This minor decrease, while statistically significant, highlights the fluctuating nature of labor market dynamics in India and the challenges in achieving substantial reductions in unemployment. The slight reduction in the unemployment rate must be analyzed in the context of India's broader economic policies, labor market reforms, and the global economic environment. While any decrease in unemployment is generally viewed positively, the practical implications of such a minor change are limited in terms of improving the overall employment situation in the country.

In the social sciences, for example, a study might find a statistically significant difference in the level of happiness between two groups of people after a certain intervention (and after eliminating the relevant confounders). The MID concept can help determine whether this difference is *practically meaningful*. It might be that the

[1] 04 January 2024 (CMIE), https://www.cmie.com/kommon/bin/sr.php?kall=warticle&dt=20240104181058&msec=536

difference, although statistically significant, does not translate into a *noticeable change* in the individuals' QoL.

Moreover, what is usually not considered is that there are both positive and negative confounders that influence outcomes such as salary rise, a new relationship, buying a new car or house (the positive ones), a disease in the family, a sudden death, health status deterioration (the negative ones), etc., that influence outcomes. These so-called confounders (or extraneous variables) are usually not considered in studies like the one by J.D. Power. However, I must admit that given the very large sample size, these subtle effects are negligible.

Other examples could be *voting patterns in an election result*. Similarly, *slight shifts in public opinion can lead to substantial social reforms*. Another confounder can be in this context if we learn about the public opinion before the elections. This might, of course, influence our vote for this or the other party. Similarly, first voters or the undecided ones can be influence by their parents or their peers, an effect that we cannot control.

Within education, the same can be said of the MID concept in regard to the adoption of teaching methods or changing educational practices. For example, when implementing a new teaching method, a teacher might want to find out the lowest level of improvement in student performance that is notable. In standardized testing, some teachers may set benchmarks whereby, once a student achieves the required points, they will change their approach to teaching. Indicates a need for the adoption of a MID approach. I stress again that the need for the adoption of a MID approach in such scenarios too.

In addition, I advocate the view that each student has to be treated differently: a three-point on a ten-point scale increase for a topper might be a disappointment but at the same time it can be a glimpse of hope and success for an under-achiever. I firmly believe that my hypothetical example also speaks for an individualized MID not just in the clinical field.

For instance, test score comparability often causes problems for teachers trying to make sense of score variances. One piece of research looking for the effects of a new education design on students' learning outcomes may conclude that indeed there is an improvement; however, the magnitude of change in scores may not be large enough to cross the MID threshold. This underscores the importance of having MID in educational research to ensure that findings will lead to actionable strategies that will enhance student outcomes.

In this manner, policies can be put in place knowing in advance what the benefits of the action will be and how they will impact future policies. The establishment of an MID for efforts indicators will assist researchers in appreciating what the policy shifts mean and whether or not it matters to those affected by the policy changes.

Moreover, the MID approach may also be applied to the evaluation of technology products and services. Businesses in the tech world often aim at improving UX (user experience) and make design changes in a stepwise manner. Having the MID in place allows firms to pinpoint which of the improvements in user satisfaction are actually relevant to their customers.

To illustrate, a software development firm can create a user survey and analyze what the users consider as a balance of usability on the current interface design. In this case, a guarantee that the changes meet the user's needs can be assumed.

The idea of MID is especially germane in the world of technology at the level of user experience (UX) design. To understand how satisfied users are, researchers who specialize in UX tend to utilize a variety of metrics, and establishing an MID enables him or her to focus on changes which will have the greatest effect on user participation. For example, an update in the appearance of a website might statistically show a lower level of user abandonment; however, if the increase is not above the MID level, it may not be important; one can put it down to random statistical fluctuation. This illustrates the need to apply MID in UX design research in order to ensure that the design choices made are based on what users consider valuable.

Application of the MID approach can be relevant in public health when evaluating health intervention for a community. For instance, during an evaluation of a new public health anti-smoking initiative, the research questions may include how much smoking prevalence needs to be reduced for the community to consider it significant in the sense of a noticeable change in satisfaction, hospitalizations, reduction of chronic obstructive pulmonary disease (COPD) cases, etc. I strongly argue that it *should be* through this method that health officials determine how effective their programs are, and what changes need to be made to increase health outcomes. Understanding the perceptions of communities toward health interventions through the set MID of health outcomes enables changes toward appreciation and acceptance of the initiated positive behavior change.

Moreover, within the domain of environmental studies, the MID approach can be relevant. For instance, researchers who are investigating the effects of environmental policy on air quality may seek to find out what the most minimal enhancement of collective air quality indices is accepted by the target communities. This will give information to the political stakeholders regarding the success of the initiatives and how that should influence the next environmental policies. The same approach ensures that the policies put in place are not counter to the expectations of the community.

Most importantly, without relevant patient characteristics, there is no way of determining the MCID with confidence (Johnston et al., 2015). To cite Johnston et al. (2015): "Without individual patient data, we will not be able to explore, for example, subgroups of patients with mild or severe disease, gender differences and the relative contributions of these factors. We will alert our knowledge users to this potential limitation." Their standpoint is in accordance with my view of an individualized MCID/MID.

When examining sports and physical endeavors, moderate amounts of improvement do work to evaluate the results of training programs or focuses targeting the improvement of an athlete's effectiveness. Coaches and trainers may want to pinpoint the smallest measure of improvement that is regarded as beneficial, for example, faster sprint times (measured in milliseconds). When switching into more practical application of MID using in sports science research, practitioners are able to better adapt their intervention to the requirements of the athletes.

MID can also be applied in *marketing*. For instance, it is possible to measure the effectiveness of an advertising campaign by measuring the changes in the *reaction of the target group*. Marketers may want to examine the smallest increment of increase in brand recognition or intention to buy something that is viewed as beneficial. This kind of information is useful in formulating enhanced marketing decisions and in managing marketing efforts. If MID can be defined for marketing outcomes, it is easier for marketers to understand their targeted consumers and how well their campaigns perform.

In social interventions, the MID is used to analyze the parallels with the social compilations. The idea of the employment of MID in settings is to find improvements where participants perceive it as beneficial in life standards. This enables better social understanding regarding the efficacy of social programs, and how they should be in the future—the better understanding of the effectiveness of social programs issue. The application of MID in practice in social sciences aids in correlating the work of practitioners with the actual needs and perceptions of the social groups that are served.

MID in Litigation

There is an avenue for an uncharted area for MID, namely, that it can also be extrapolated into criminal litigations, especially in estimating psychological harm or the success achieved through rehabilitation programs, for example. There are a few legal cases where MID has been used in criminal litigation matters. I have not found any, actually. However, when applied in legal matters, particularly those with respect to psychological evaluation, restorative justice, or the psychosocial effects inflicted on the victim or the offender, it can be a useful tool.

Experts can argue with MIDs before the judge to determine the extent of damage and also longitudinally in civil law. Tort, which word comes from Latin, refers to the wrongful act to another person (not from the field of criminal law). In essence, tort provides remedies. It is usually monetary compensation for the injured individual. There are different types of torts: personal injury, product liability, defamation, and invasion of privacy. The question in this context arises how to determine the monetary compensation. I contend that MID should be used and administered by an expert.

Bouffard et al. (2012) note that restorative justice (RJ) plays an important role in reducing the rate of reoffending in juvenile offenders. It follows, therefore, that the evaluation of effectiveness of these interventions must also be done from the perspective of MID, in my opinion, to establish the levels of change in recidivism and victimization. This way, some lower level threshold for the outcome of restorative justice programs can be set and therefore have an impact on the legal decision-making in criminal cases. To exemplify the intervention that they implemented, Bouffard et al. (2012) used mediation, and also victim/offender meditation.

My hypothetical study would examine the effectiveness of restorative justice programs and suggests that assessing outcomes through the lens of "MID" can enhance the understanding of meaningful change in recidivism rates. To simplify it, how many months of detainment would suffice, so that the perpetrator will not commit the same crime or another crime again.

In addition, the application of MID can be salient in circumstances where there are psychological evaluations aimed at assessing the effects of crime on the victims. The mental injury that the victims go through can be measured and evaluated on a standardized scale in order to establish whether the measures taken such as counseling and restorative meetings were useful in improving the mental health of the victims.

The work by Bouffard et al. (2012) is largely consistent with the findings of Wemmers and Cyr (2004), who look into the views of the victims of restorative justice and how the outcomes from the processes of restorative justice should be negative (Wemmers & Cyr, 2004).

Wemmers and Cyr (2004) show findings from a study focusing on victims of crime invited to partake in mediation sessions with their offenders. In addition, their paper also discusses the procedural justice theory which outlines the rationale behind a victim's role preference in the criminal justice system. The paper ends with suggestions to adopting a criminal justice system centered toward victims.

In my opinion, the inclusion of MID in these analyses may help in assessing how well restorative justice works relative to the pain inflicted on the victims. Also, in my mind, both advocates and prosecutors can also argue with medical expert opinions along with MID as supporting evidence.

Kemi (2024) explains the historical background of plea bargaining in the American legal system, claiming that plea bargaining was invented in order to make legal processes simpler yet effective in serving justice (Kemi, 2024). The negotiation where plea bargains are made considers the psychological and social ramifications of the crime on its victims and its perpetrators. The framing of MID could help in assessing if such agreements would be useful and meaningful for all participants. Plea bargaining could be one of the first historical cases of the roots of MID. By the way, Kemi (2024) found that judicial decisions in both Nigeria and Kenya demonstrate that plea bargaining serves as an effective instrument in combating corruption.

To sum up, although there are few if any specific cases of the use of MID in criminal litigations, its concepts can be useful and even applicable in many, particularly in restorative justice, plea bargain, and assessment of the psychological aspects of victims. The fact that MID can be integrated into the frameworks and contexts identified is critical because it expands the understanding of what is a valuable change and improvement in the criminal justice system, and changes that are more valuable to all involved.

References

Bouffard, J., Cooper, M., & Bergseth, K. J. (2012). The effectiveness of various restorative justice interventions on recidivism outcomes among juvenile offenders. *Youth Violence and Juvenile Justice, 10*(3), 275–290. https://doi.org/10.1177/1541204016647428

Fahmi, A., Mukminatien, N., Ginting, D., & Kusumaningrum, S. R. (2024). The impact of flipping class intervention on reading comprehension: Different approaches and proficiency levels. *PLoS One, 19*(6), e0305041. https://doi.org/10.1371/journal.pone.0305041

Jaeschke, R., Singer, J., & Guyatt, G. H. (1989). Measurement of health status: Ascertaining the minimal clinically important difference. *Controlled Clinical Trials, 10*(4), 407–415. https://doi.org/10.1016/0197-2456(89)90005-6

Johnston, B. C., Ebrahim, S., Carrasco-Labra, A., Furukawa, T. A., Patrick, D. L., Crawford, M. W., Hemmelgarn, B. R., Schunemann, H. J., Guyatt, G. H., & Nesrallah, G. (2015). Minimally important difference estimates and methods: A protocol. *BMJ Open, 5*(10), e007953. https://doi.org/10.1136/bmjopen-2015-007953

Kemi, O. (2024). Assessing the implication of plea bargain under ACJA, 2015 in Nigeria's anti-corruption crusade: Lessons from Kenya. *Asian Research Journal of Arts & Social Sciences, 22*(1), 13–29.

Norman, G. R., Sloan, J. A., & Wyrwich, K. W. (2003). Interpretation of changes in health-related quality of life: The remarkable universality of half a standard deviation. *Medical Care, 41*(5), 582–592. https://doi.org/10.1097/01.MLR.0000062554.74615.4C

Power, J. D. (2025, May 7). *2025 North America Airline satisfaction study*. Retrieved from https://www.jdpower.com/business/press-releases/2025-north-america-airline-satisfaction-study

Stone Fish, L., & Busby, D. M. (2005). The Delphi method. In D. H. Sprenkle & F. P. Piercy (Eds.), *Research methods in family therapy* (2nd ed., pp. 238–253). The Guilford Press.

Wemmers, J.-A., & Cyr, K. (2004). Victims' perspectives on restorative justice: How much involvement are victims looking for? *International Review of Victimology, 11*(2), 145–161. https://doi.org/10.1177/026975800401100204

Zhang, Y., Xi, X., & Huang, Y. (2023). The anchor design of anchor-based method to determine the minimal clinically important difference: A systematic review. *Health and Quality of Life Outcomes, 21*, 74. https://doi.org/10.1186/s12955-023-02157-3

Chapter 9
MID in Ecology and Environmental Sciences, Land Use Change, and Soil Health

Abstract The minimal important difference (MID) is a concept that is known within clinical and psychological research, but it has recently gained its most important implications in the environmental and ecological sciences. In these areas, MID is defined as the smallest change that can be measured in an environmental variable that impacts ecosystem functions or services in a quantifiable manner. These changes could be subthreshold, such as a small temperature increase or a minor reduction in precipitation. However, if the change surpasses a threshold, it can affect the ecological system.

Ecosystems are complex, multivariate systems that display a great deal of sensitivity and offer essential ecosystem services. Perturbations caused by anthropogenic activities such as urban expansion, deforestation, pollution, the introduction of non-native species, or the overutilization of natural resources are bound to change plant–animal interaction networks, hydrological cycles, and much more. It is within the scope of these changes that the MID can be used to verify the extent of damage or, conversely, the progress achieved.

For example, forest restoration initiatives exemplify the potential usefulness of MID by enabling guided monitoring through measurable indicators like increased tree population, animal recolonization, and enhanced soil quality.

Likewise, in marine ecosystems, the MID helps assess whether shifts in biodiversity or water quality warrant ecological attention. Nevertheless, there are still issues related to high variability, unpredictable natural phenomena, and human impact deviations when attempting to define MIDs in other than clinical environments. These issues, however, can be solved with strong statistical modeling.

In studies of soil health and land use, with MID, critical thresholds in soil characteristics, such as organic carbon, water retention, and microbial activity, which decline due to poor land management, can be systematically defined. In the Himalayan region and Ireland, studies have documented the impact of land cover and agricultural practice changes on soil erosion, structure, and fertility. In these cases, MID serves as a decision-support system for assessing the effectiveness of various soil conservation strategies, including organic farming and crop rotation. MID strengthens participation and commitment by defining what change represents

I. Fekete, *Transforming Data into Informed Decisions across Clinical and Non-Clinical Domains*, https://doi.org/10.1007/978-3-032-02595-1_9

an ecological meaningfully achievement, thus fostering responsible environmental management.

In the final chapters, it is advocated for a move from generic estimations of MIDs to tailored models for ecosystems, individuals, or biological systems. Incorporating appropriate variables into predictive models can enhance the precision and usefulness of MID calculations. This approach to MID is meant to have a paradigm-changing impact on ecological and environmental research, much like its application in medicine and psychology.

Keywords MID in ecology and environmental sciences · MID in forest restoration · MID in marine ecosystem restoration · Impact of MID-based restoration · Ecological restoration · Biodiversity · Conservation · MID in land use change and soil health · MIDs integration into policy regulations · Sustainability

MID in Ecology and Environmental Sciences

Analogous to the MCID, there is an ecological parameter that would result in a meaningful change in the ecosystem (Groffman et al., 2006). This could be a slight increase in temperature, or a minor decrease in rainfall, or a little shift in the population of a particular species. These statistically significant changes, when they cross the minimal important difference (MID), can trigger a vast effect that may change the balance of the whole ecosystem.

These changes often have great impacts on ecosystem functionality and the services they provide to human societies. This is why MID is important in the "eco-domain" too. For example, rapid shifts in climate conditions can lead to disturbances in ecosystem services. Fontúrbel et al. (2021) warn that climate change may have an effect on individual species. Moreover, their ecological interactions between plants and animals can also be disturbed, ultimately leading to plant mortality and the reduction of the production of fruits.

Another example is that increased temperatures can have an effect on *water supply* and lead to changes in the *distribution of species* and increase the *frequency and intensity of wildfires and pest outbreaks* [The Intergovernmental Panel on Climate Change (IPCC), 2021]. Second, urbanization, deforestation, and conversion of land for agriculture or industrial use can lead to the loss of ecosystem services such as flood regulation, air and water purification, and carbon sequestration. For example, deforestation in tropical areas greatly affects global and local climates and reduces biodiversity.

Third, the introduction of non-native species can alter local ecosystems. These species can outcompete native flora and fauna, alter habitats, and disrupt the provisioning of ecosystem services like freshwater availability and, again, water quality (Pejchar & Mooney, 2009). Fourth, chemical, physical, and biological pollutants can degrade ecosystems, not just our health. For instance, nutrient loss can cause hypoxic conditions, thereby killing fish and other aquatic animals (Carpenter et al., 1998).

This in turn affects food provision and water quality, as already mentioned before (Carpenter et al., 1998). Finally, unsustainable harvesting of resources, whether it be logging, fishing, or water extraction, can lead to a rapid decline in ecosystem capacity to renew these resources.

For example, Groffman et al. (2006) examined ecological thresholds in *urban ecosystems*, offering insights into how small changes in *urban land use* or *pollutant loads* can significantly *impact urban water quality and biodiversity*. Their research underscores the importance of identifying and managing ecological thresholds in order to prevent abrupt and undesirable changes in ecosystem services. Their study discusses the *practical application of ecological thresholds*, analogous to MIDs, in *environmental management*, emphasizing their importance in identifying meaningful ecological changes.

Suding and Hobbs (2009) looked at how important models are for managing places where animals and plants live. So far, research has shown that human actions can increase the chances of these sudden changes happening in more places and cause ecosystems to change into new states that are hard to fix. Suding and Hobbs (2009) propose a way to use *threshold models for habitat management*, similar to MIDs, that focuses on making them useful for *quick decisions* and managing areas affected by humans. This method highlights the importance of using these models carefully, making sure that they work well and are relevant for solving today's environmental problems.

In a similar vein, Suding and Hobbs (2009) reflect on ecological restoration, emphasizing the need for setting clear objectives and measurable targets that signify meaningful ecological recovery, analogous to MID concepts. Suding et al. (2015) evaluate the success of interventions aimed at restoring degraded ecosystems. An example they bring explores the *ecological thresholds* and *minimal changes in habitat conditions* necessary to *achieve significant improvements in ecosystem health and biodiversity*.

However, the challenge lies in accurately determining the MID in different ecological domains. It is further complicated by the inherent variability and unpredictability of natural systems, and a bunch of factors that can influence ecological processes.

MID in Forest Restoration

When we talk about the restoration of forests, it means reintroducing indigenous species, increasing biodiversity, and improving the services provided by the ecosystem. Given this, there are many factors that need to be considered, which makes it difficult to ascertain the success of these efforts in such situations. This is where the notion of MID becomes relevant.

This case presents a scenario where restoration of forests was attempted in a region with forests in an ailing condition. To facilitate monitoring of progress in restoration, several indicators were established, such as the increase in population of the indigenous trees, the emergence of animal species, and the enhancement of soil parameters. The difficulty was posed by the issue of the threshold of these parameters.

With regard to the particular case study, the application of MID proved to be useful in making sense of the success of restoration activities as a whole. Instead of relying on arbitrary or absolute values, the MID allowed for a more context-specific and relative evaluation of forest restoration.

These lessons from this case study are not limited to the case of forest restoration; they are broader. Regardless of the sector, be it health care, education, or environmental conservation, the MID is a powerful technique for capturing and understanding substantial/meaningful change.

MID in Marine Ecosystem Restoration

The marine ecosystem is a complex, dynamic system that is home to a diverse array of species. It plays a crucial role in regulating the planet's climate and providing food and livelihood for millions of people. However, human activities such as overfishing, pollution, and climate change have caused substantial damage to these ecosystems. The task of restoring them is immense.

Applying MID in this scenario can help in quantifying the changes needed to bring about a significant improvement in the health of the marine ecosystem. We have to focus on a variety of indicators such as water quality, biodiversity, and the population of key species.

However, employing MID has some challenges in these nonclinical fields too. For example, findings could be impacted by weather conditions, migration of species, and even artificial human interventions. Therefore, advanced statistics should handle these uncertainties.

Analyzing the Impact of MID-Based Restoration

In the context of restoration, MID can be seen as a tool that enables us to measure the success of our efforts objectively. It provides a benchmark, a quantifiable standard against which we can assess the effectiveness of our strategies. It allows us to move beyond subjective evaluations and toward a more evidence-based approach.

However, as with any tool, the use of MID is not without its challenges. One of the key issues lies in defining what constitutes an important change. This is not always a straightforward task and can be subject to a range of factors, including individual subjective perceptions and expectations. Furthermore, the concept of MID assumes that any change above the threshold is beneficial, which may not always be the case.

Despite these challenges, MID-based restoration has significant potential. It offers a way to bring greater objectivity and rigor to the process of restoration, enabling us to measure our efforts more accurately and effectively. It also encourages us to think critically about what constitutes "success" in restoration and to strive for more than just the minimum.

Environmental sciences encompass a broad range of disciplines that study the interactions between the physical, chemical, and biological components of the environment. The integration of these disciplines is crucial for understanding complex environmental issues such as climate change, pollution, and biodiversity loss. MID could provide a framework for determining the smallest change in environmental indicators that can be considered meaningful from a policy perspective.

For instance, when evaluating the impact of a new conservation strategy, researchers can utilize MID to ascertain whether observed changes in species populations or ecosystem health are significant enough to warrant continued investment in that strategy.

The application of MID in environmental sciences can also enhance the design of studies aimed at assessing the effectiveness of interventions. For example, in studies evaluating the impact of pollution control measures, establishing a clear MID for air quality improvements can help policymakers make sense of the real-world implications of their decisions (Silva & Torres, 2022).

By focusing on changes that exceed the MID, researchers can provide more robust evidence to publicly support the implementation of environmental regulations. This approach not only aids in the scientific evaluation of interventions but also facilitates communication with stakeholders, including policymakers and, last but not least, the public, about the significance of environmental changes.

Moreover, MID can be instrumental in the field of environmental monitoring. As environmental data become increasingly complex and voluminous, establishing MIDs for various indicators can help prioritize which changes are most critical to monitor. For instance, in assessing the health of aquatic ecosystems, researchers may define MIDs for key indicators such as nutrient levels or species diversity. This allows for a more focused approach to monitoring efforts, ensuring that resources are allocated effectively to track changes that matter most for ecosystem health (Silva & Torres, 2022).

Ecological Restoration

Ecological restoration aims to return degraded ecosystems to a stable and healthy state. The role of MID in ecological restoration is particularly important, as it provides a benchmark for evaluating the success of restoration efforts. By defining MIDs for various ecological indicators—such as species richness, vegetation cover, or soil health—restoration practitioners can assess whether their interventions are achieving meaningful outcomes. For example, in a restoration project aimed at reestablishing native plant communities, a defined MID for plant cover can help determine whether the project is successful in enhancing biodiversity and ecosystem function.

Restoration practitioners can use MIDs to set realistic goals and expectations for their projects, ensuring a focus on achieving outcomes that are not only statistically significant but also ecologically meaningful. This is particularly important in large-scale restoration projects, where the complexity and variability of ecosystems can make it challenging to interpret results. By establishing clear (and multiple) MIDs, practitioners can better communicate the significance of their findings to stakeholders and secure ongoing support for restoration initiatives (Blair & Houghton, 2019).

Furthermore, the integration of MID into ecological restoration can enhance adaptive management practices. By continuously monitoring ecological indicators against established MIDs, practitioners can make informed adjustments to their restoration strategies based on real-time feedback. This iterative approach allows for more effective management of restoration projects, ultimately leading to improved outcomes for ecosystems and the services they provide.

Biodiversity and Conservation

Biodiversity conservation is critical for maintaining ecosystem resilience and functionality. The concept of MID is particularly relevant in biodiversity assessments, where it can help determine the significance of changes in species populations or habitat conditions. For instance, when evaluating the effectiveness of conservation strategies, establishing MIDs for species abundance or diversity can provide a clearer picture of whether conservation efforts are making a tangible impact. This is essential for justifying the allocation of resources to specific conservation initiatives and for demonstrating accountability to stakeholders.

In biodiversity conservation, MID can also play a role in the design of monitoring programs. By defining MIDs for key biodiversity indicators, conservationists can prioritize monitoring efforts and focus on changes that are most likely to reflect the success of their interventions.

For example, in a project aimed at protecting endangered species, establishing MIDs for population growth rates can help determine whether conservation actions are effective in promoting recovery. This targeted approach to monitoring can enhance the efficiency of conservation programs and ensure that efforts are directed toward the most impactful strategies (Blair & Houghton, 2019).

By establishing clear and meaningful benchmarks for the success of biodiversity, conservationists can engage stakeholders in discussions about the significance of biodiversity changes and the importance of continued conservation efforts (Silva & Torres, 2022; Kim et al., 2016). This collaborative approach can foster a sense of trust and shared responsibility for biodiversity conservation and encourage collective action to protect ecosystems: collective action among stakeholders, including government agencies, nonprofit organizations, and local communities (Silva & Torres, 2022; Kim et al., 2016).

The Concept of MID in Land Use Change and Soil Health

To start with, Dörner et al. (2016) claim that poorly managed land use changes can cause soil imbalances that negatively impact soil properties like soil depth which is important for storing moisture and nutrients. This clearly shows that using MIDs to track changes is essential in this field too.

Dörner et al. (2016) also demonstrate that the soils are shallow because of the existence of an impermeable stratum. This in turn causes water logging for several months during the year. These fragile soils, known as "Ñadi," have experienced some form of intensive land use change (that resulted in a reduction in soil depth). The focus of their research was to study the effect that the land use change in Ñadi soils had on their physical properties over time and space.

Changes in the water-table depth and penetration resistance over time and space were observed too. Dörner et al. (2016) used sensors located in three depths, which continuously measured volumetric water content and soil temperature, and these depths also had their water retention curve and air permeability measured at those depths. Unfortunately, findings revealed that the changes in land use over time have resulted in lower soil depth. The soils beneath naturalized grassland *exhibited lower values for soil water storage capacity*, *air capacity*, and *permeability* when compared to the soils beneath secondary naturalized grassland.

In addition, MID can assist in identifying particular indicators or markers of soil health that are impacted by land use changes. The evaluation of soil organic carbon (SOC), as stated by Don et al. (2010), reveals that changes in land cover modulate SOC stocks, potentially altering the ability of the soil to sustain agricultural productivity and ecosystem services. By employing MIDs in soil health assessments, the stakeholders might grasp the effect land use choices have on soil health and sustainability better.

The Role of MID in Predicting Evolving Dynamics of Soil Properties Brought by Changes in Land Use

Although not called MID, MID is very useful in assessing the significant changes in soil health related to land use changes. For instance, according to Buckingham et al. (2014), it is highly important to monitor changes in soil carbon stocks [soil organic carbon (SOC)] to comprehend the long-term effects of land use changes, especially in the pregnancy of climatic change and agriculture (Buckingham et al., 2014). SOC is very important in that it supports the storage of CO_2. The successive use of MID could help recognize anomalies in soil health parameters, which can later help in taking preventive measures. This is exactly what Buckingham et al. (2014) suggest.

Moreover, MID can also be very beneficial in measuring the impact of various land use management techniques. Changes in land use can be less harmful to soil health in the case where sustainable agriculture practices are followed, for example, in Switzerland. I am referring to the Zurich Organic Fertilization Experiment (ZOFE) in Switzerland.

Similarly to the study by Buckingham et al. (2014) and Borrelli et al. (2017) contend that the adoption of sustainable agricultural practices can greatly reduce soil erosion caused by land use by humans, and increase phosphorus recycling, which improves soil health. Again, the application of MID would help the scientists and users of land evaluate the effectiveness of these practices and modify the policies.

In my mind, there must be rigid limits; however, I think given the dynamic nature of change in the ecosystem, these limits might change over time too. My statement is supported by Borrelli et al. (2017): their estimation of 35.9 Pg year-1 of soil eroded in 2012 is at least two times lower than the previously reported annual soil erosion reference values.

Moreover, MID facilitates the assessment of different land utilization systems based on their soil health indicators. For example, Saha and Kukal (2013) explain that regarding the various land use systems, it is important to understand their impact on soil if changed or adapted for the context of environmental sustainability. MID provides selected stakeholders with the ability to determine specific practices that encourage these changes and promote them more widely.

In recent years, the greater Himalayan region of northwest India experienced a transformative alteration in land use (Saha & Kukal, 2013). This, therefore, called for scientific investigation to understand the effect of grasslands, forests, agriculture lands and eroded lands on soil erosion, soil aggregation, water retention, etc.

The soil samples were assessed. The surface layers of grassland exhibited the greatest water-stable aggregates (WSA[1]) >2 mm of 17.3%, while the microaggregates (WSA <0.25 mm) were found to be the greatest in eroded soils (Saha & Kukal, 2013). Water drop stability showed an analogous pattern. It declined with the increase in size of the aggregate. Eroded soils had the lowest soil organic carbon, which means that it suffered the brunt of land use change over time.

The effects of land use, soil depth, and their interaction was significant ($p < 0.05$) in the water retention at varying soil water suction levels. Eroded soils had lower water retention in contrast to grassland and forest soils, which were significant ($p < 0.05$) as per Saha and Kukal (2013). Eroded soils showed a noticeable decline in both saturated hydraulic conductivity and maximum water holding capacity relative to the forest and grassland soils. This was indicative of the deterioration of soil physical properties as a result of natural ecosystem farming system conversion coupled with increasing erosion in the northwestern lower Himalayan region of India (Saha & Kukal, 2013). I have brought these examples just to show that MID literally is related to the whole globe including the soil and the animals not just our health and air-quality.

[1] Water-stable aggregates.

The Common Indicators of Soil Health and MID

Soil indicators that are used more regularly include soil organic matter, nutrients, and microbial content. Such indicators are important for assessing whether and how land utilization practices change, and also their effects in comparison to the previous use. MIDs are important within the context because they provide the guidance that is needed for the selection and monitoring of these indicators as a systematic process.

For instance, soil carbon stocks have been shown by Oberholzer et al. (2014) to change when land use changes, in line with the previously mentioned studies such as Buckingham et al. (2014), Borrelli et al. (2017), and Saha and Kukal (2013).

Thus, agricultural SOC should be monitored as an indicator of soil health as per Oberholzer et al. (2014). With MIDs, it is possible to set proper limits or targets for the monitoring indicators, which help in assessing whether the practices are acceptable or not, as already mentioned (e.g., by using sensors, see Dörner et al., 2016).

Let me elaborate on the study by Oberholzer et al. (2014). They claim that converting grassland into farmland usually reduces the amount of organic carbon stored in the soil and lowers soil quality. However, using fertilizers, rotating crops, and managing plant leftovers can help reduce or even reverse these losses. Their study looks at soil carbon and crop yields over 60 years in a Swiss farming experiment (called ZOFE) that I already referred to at the beginning of this chapter. It tested 12 different ways of fertilizing the soil, using both natural and chemical fertilizers.

Since the land was turned into farmland in 1949, all treatments have shown ongoing soil carbon loss, at a rate of 0.10–0.25 tons of carbon per hectare per year (Oberholzer et al., 2014). The carbon being added back through fertilizers and plant materials ranges from 0.6 to 2.4 tons per hectare per year, but it is not enough to stop the losses. The soil still has not stabilized to a new balance.

Crop yields improved over the years due to better fertilizers and plant breeding, but high yields were only possible when chemical fertilizers were used at modern levels (Oberholzer et al., 2014). If only natural fertilizer was used without extra minerals, yields dropped significantly—by 25–50%, exceeding the MID (remark mine).

Overall, the study by Oberholzer et al. (2014) shows that turning grassland into farmland leads to major losses in soil carbon—between 15% and 40% (exceeding the MID, again, remark mine)—even when farmers use advanced, eco-friendly techniques.

In addition, the use of MID in measuring the soil health indicators will improve the relationships between soil characteristics and land use management. Keith et al. (2012), for example, indicate that land use practices are important determinants of soil fauna diversity, which are critical for soil health, in accord with the studies cited so far. With the aid of MIDs, researchers would be able to explain such relationships and propose strategies for the improvement of soil health, I should think.

Keith et al. (2012) focused on how to develop reliable bioindicators for assessing soil quality. They examined the types of soil organisms such as bacteria, fungi, mycorrhiza, nematodes, mites, earthworms, and ants in Ireland. including farmland,

Table 9.1 The forest land has substantially higher soil fertility indicators, this table compares soil chemical properties across different land uses (farm, forest, and grazing) at two soil depths (0–20 cm and 20–40 cm), with forest land having the highest beneficial values almost all around

Soil depth (0–20 cm)								
LLand use	pH	% SOC	%TN	AvP	EEx.Ca	EEx.Mg	EEx.K	CEC
FFarm land	4.63^{c}	1.33^{e}	0.11^{e}	12.63^{c}	9.27^{d}	5.83^{c}	1.13^{b}	25.20^{b}
FForest land	6.02^{a}	3.79^{a}	0.35^{a}	27.83^{a}	23.30^{a}	12.70^{a}	1.51^{a}	37.55^{a}
GGrazing land	5.14^{b}	2.29^{c}	0.20^{c}	4.17^{e}	14.43^{b}	6.77^{b}	1.04^{b}	27.31^{c}
Soil depth (20–40 cm)								
FFarm land	5.20^{b}	1.19^{f}	0.10^{f}	10.43^{d}	6.87^{e}	4.20^{a}	0.88^{cd}	24.04^{b}
FForest land	6.02^{a}	3.10^{a}	0.28^{b}	24.52^{b}	12.27^{c}	6.17^{b}	0.79^{c}	24.10^{c}
GGrazing land	5.13^{b}	1.70^{d}	0.15^{d}	2.43^{f}	8.17^{a}	5.47^{a}	0.44^{e}	23.86^{b}
Pp(0.05)	00.03	00.05	00.01	00.05	00.02	00.04	00.016	00.045

Original table from Welemariam et al. (2021). (In their article it is Table 4)
Note. The superscripts "a, b, c, and d" denote statistically significantly different results

pastures, forests, and bogs. They also looked at how these organisms relate to soil characteristics such as its physical and chemical properties.

Importantly, Keith et al. (2012) discovered that the presence and variety of nematodes and earthworms were good indicators of the overall soil community. Using species that strongly indicated specific soil types or conditions, the researchers identified certain organisms that could serve as effective bioindicators.

Overall, Keith et al. (2012) show that different types of soil life can act as surrogates for one another when assessing soil health, but the best indicators might differ between heavily farmed areas and more natural landscapes. Accordingly, MIDs should change, if I may add.

Moreover, it is possible to capture soil health indicators and the differences between regions with the help of MIDs. For example, land use practices are known to affect soil carbon stocks in different places, with researchers such as Welemariam et al. (2021), conducting their studies in Ethiopia, reported on a lot of values. However, MIDs could have come in handy when determining relevant differences (Table 9.1).

Welemariam et al. (2021) should or could have used MIDs to foster their results, given the minute differences. The derived evidence from using MID could guide the policymaker to develop appropriate land management measures targeted to improve soil health.

Regional Examples, Best Practices Toward Sustainable Land Use and in Soil Management, and Precise Guidelines for Policymakers

Successful practices of land use at a mid-scale level should integrate MID into the framework of sustainable land management, in my opinion. In the Loess Plateau of China, for example, the large-scale restoration of ecosystems has been undertaken by guiding sustainable land management efforts in the region demonstrating MID works (Ren et al., 2020). Also successful is the Grain-for-Green Program, as it has served other regions of the world needing afforestation and soil erosion control.

The combination of conservation agriculture, agroforestry, and organic farming all are example of best practices in soil management. Similar ones focus on conservation agriculture, agroforestry, and organic farming.

Empirical work done by Khadka et al. (2018) mentions the need of determining soil fertility and applying the necessary management practices for better soil health in Nepal. This explains why and how MID can assist farmers in depopulating the soil fertility by informing them on the necessary fertilizer to be applied and what crops to rotate with, thus improving soil health and agricultural productivity.

Policymakers should pay special attention to the incorporation of MID into the regulatory frameworks of land use planning. This ought to involve setting nominal thresholds of soil health indicators and increasing land management systems that are eco-friendly.

For example, Tesfahunegn (2013) pointed out the necessity of soil quality indicators such as soil texture, structure, pH, nutrient content, and biological activity in making land use decisions. These indicators are very helpful in determining the ability of plant growth and soil health. There is a great argument that there should be policies that encourage the use of relatively better practices, such as cover cropping or rotating crops, which are ways that have been shown to enhance soil quality. Again, MIDs could determine if some change on these indicators mentioned above are of substantial or not.

The Use of MID in Designing and Adopting Environmental Regulatory Policies and Standards: The Integration of MIDs into Policy Regulations

It is crucial to integrate MID into environmental regulations and standards in order to enhance the sustainable use of land. As is pointed out by Alawamy et al. (2021), it is indeed necessary to develop good land-use management strategies to ensure that soil will be there able to sustain productivity in the future. When it is finally done, we will be more confident that incorporating MID into policy frameworks is one of the effective ways of tackling soil degradation and unsustainable land management.

In addition, MID may enhance the design of adaptive management approaches as a response to the changing state of the environment. Indeed, as pointed out by Guo (2024), appreciating how land use change alters soil erosion is crucial for making policies. The use of MID in integrated environmental assessments enables policymakers to formulate regulations that are open to modification using up-to-date information and new pieces of research.

In addition, the contribution of MID in setting minimum requirements goes beyond national boundaries and international cooperation. More than ever, the global community appreciates the necessity of preserving land resources as a means to attain the sustainable development goals (SDGs). MID should be incorporated into international standards so that various countries will be able to meet their sustainability commitments.

Making a Case for Policymakers, Researchers and Practitioners on Ensuring Soil Health as a Critical Factor for Environmental Responsibility and Agriculture Productivity

Land use changes that compromise soil health pose major food and ecosystem sustainability challenges. As Panday et al. (2018) noted, determining the variability of soil characteristics is important for good management practices. There is a need for the engagement of policy formulation, research, and practice to consider soil health as part of their action plans.

A mobilization strategy should help to promote the significance of soil health as it touches on issues of sustainable development. As such, research geared toward the education of farmers and land managers should also be promoted and examine the effects of land use changes on different soil properties.

According to Willy et al. (2019), we should adopt an integrated approach to sustainable land management because of the malpractice of farmers. They suggest that to improve soil fertility and reduce the adverse effects of exogenous agricultural activities (Willy et al., 2019). However, they also add that the deterioration is not just due to human maltreatment of the soil but also to "inherent soil properties" (Willy et al., 2019). Through such collaboration it will be possible to achieve protection and restoration of soil health resources by stakeholders.

Hundreds of examples demonstrate that information on the impact of land-use systems on underlying soil properties is critical to effective management practices (Bizuhoraho et al., 2018).

However, Bizuhoraho et al. (2018) could have used MIDs along with their statistical analyses. Besides, their statistics are wrong: they report on a *p*-value of 0.171 (EP[2] cmol/kg), which "becomes" 0.026 (LSD), see p. 36 (Table 1 in Bizuhoraho et al., 2018, p. 36). In Table 1 in Bizuhoraho et al. (2018), the three values for EP[3] cmol/kg are: 0.333 (farmland), 0.353 (forestland), and 0.353 (cultivated land). Also, in the same table (Table 1, p. 36), an LSD-value of 70.3 for AP ppm is highly unlikely. LSD is a *p*-value. How can it be 70.3?

---

[2] Exchangeable potassium.

[3] Exchangeable potassium.

Last but not least, the use of MID has the potential to provide a good starting point in understanding and tackling statistical issues like these. It is possible for policymakers and land managers to elevate the health of soil within the ecosystems and the agricultural systems to achieve greater sustainability and resilience of the agricultural systems. But, most importantly, MID primarily should ensure that we unravel a relevant difference, not just a random statistical fluctuation or a scientifically irrelevant but significant difference.

Let me finish by giving my opinion on MID/MCID across all fields. I have advocated throughout the book for a personalized/individualized MID/MCID. I have brought some plausible arguments supported by others too. I should emphasize that this book does not end here. I advocate a model which incorporates every relevant variable that characterizes anything, be it a human being, an animal, the soil, or even a biosensor.

Allow me to give an example from the medical field: breast cancer. Assume we compute the MCID in an RCT using the most widespread distribution-based MCID (0.5 SD). Yet, there are a bunch of other variables. My aim in the book was to introduce the MID/MCID concept in nonclinical sciences.

Now, my aim is to refine it and let the reader think further. Does the distribution-based MCID (0.5 SD) suffice? Think of a breast cancer patient. Age of the patients, tumor size in mm, tumor grade, number of positive lymph nodes, progesterone receptor in fmol, estrogen receptor in fmol, and menopausal status. All these relevant variables are not incorporated in the distributional criterion; hence, they are not incorporated in the model. A regression model could handle these variables along with the MCID in the model, a novel aspect of this book that the reader should ponder about. The dependent variable of the regression model would be the refined MCID.

References

Alawamy, J. S., et al. (2021). Response of potential indicators of soil quality to land-use and land-cover change under a Mediterranean climate in the region of Al-Jabal Al-Akhdar, Libya. *Sustainability, 14*(1), 1–19. https://www.mdpi.com/2071-1050/14/1/162/pdf

Bizuhoraho, et al. (2018). The effect of land use systems on soil properties; a case study from Rwanda. *Sustainable Agriculture Research, 7*. https://doi.org/10.5539/sar.v7n2p30

Blair, A., & Houghton, R. (2019). Data science of the natural environment: A research roadmap. *Frontiers in Environmental Science, 7*, 121. https://doi.org/10.3389/fenvs.2019.00121

Borrelli, P., Robinson, D. A., Fleischer, L. R., Lugato, E., Ballabio, C., Alewell, C., Meusburger, K., Modugno, S., Schütt, B., Ferro, V., Bagarello, V., Oost, K. V., Montanarella, L., & Panagos, P. (2017). An assessment of the global impact of 21st century land use change on soil erosion. *Nature Communications, 8*(1), 2013. https://doi.org/10.1038/s41467-017-02142-7

Buckingham, S., Rees, R. M., & Watson, C. A. (2014). Issues and pressures facing the future of soil carbon stocks with particular emphasis on Scottish soils. *The Journal of Agricultural Science, 152*, 699–715. https://doi.org/10.1017/S0021859613000300

Carpenter, S. R., Caraco, N. F., Correll, D. L., Howarth, R. W., Sharpley, A. N., & Smith, V. H. (1998). Nonpoint pollution of surface waters with phosphorus and nitrogen. *Ecological*

Applications, 8(3), 559–568. https://doi.org/10.1890/1051-0761(1998)008[0559:NPOSWW]2.0.CO;2

Don, et al. (2010). *Impact of tropical land-use change on soil organic carbon stocks – a meta-analysis*. https://doi.org/10.1111/j.1365-2486.2010.02336.x

Dörner, et al. (2016). Spatial and temporal variability of physical properties of Aquands under different land uses in southern Chile. In *Soil use and management*. https://doi.org/10.1111/sum.12286

Fontúrbel, F. E., Nespolo, R. F., Amico, G. C., & Watson, D. M. (2021). Climate change can disrupt ecological interactions in mysterious ways: Using ecological generalists to forecast community-wide effects. *Climate Change Ecology, 2*(2021), 100044. https://doi.org/10.1016/j.ecochg.2021.100044

Groffman, P. M., Baron, J. S., Blett, T., et al. (2006). Ecological thresholds: The key to successful environmental management or an important concept with no practical application? *Ecosystems, 9*, 1–13. https://doi.org/10.1007/s10021-003-0142-z

Guo. (2024). *Effect of land use change on soil loss by water erosion in the Niyang River Basin of Tibet from 1990 to 2020*. https://doi.org/10.20944/preprints202401.0140.v1

Keith, A. M., Boots, B., Hazard, C., Niechoj, R., Arroyo, J., Bending, G. D., Bolger, T., Breen, J., Clipson, N., Doohan, F. M., & Griffin, C. T. (2012). Cross-taxa congruence, indicators and environmental gradients in soils under agricultural and extensive land management. *European Journal of Soil Biology, 49*, 55–62. https://doi.org/10.1016/j.ejsobi.2011.08.002

Khadka, et al. (2018). Soil fertility assessment and mapping of regional agricultural Research Station, Parwanipur, Bara, Nepal. *Journal of Nepal Agricultural Research Council, 4*, 33–47. https://doi.org/10.3126/jnarc.v4i1.19688

Kim, H., et al. (2016). Ldb1 is essential for the development of isthmic organizer and midbrain dopaminergic neurons. *Stem Cells and Development, 25*(1), 1–10. https://doi.org/10.1089/scd.2015.0307

Oberholzer, et al. (2014). Changes in soil carbon and crop yield over 60 years in the Zurich organic fertilization experiment, following land-use change from grassland to cropland. *Journal of Plant Nutrition and Soil Science, 177*. https://doi.org/10.1002/jpln.201300385

Panday, et al. (2018). Digital soil mapping in the bara district of Nepal using kriging tool in ArcGIS. *PLoS One, 13*. https://doi.org/10.1371/journal.pone.0206350

Pejchar, L., & Mooney, H. A. (2009). Invasive species, ecosystem services and human well-being. *Trends in Ecology & Evolution, 24*(9), 497–504. https://doi.org/10.1016/j.tree.2009.03.016

Ren, et al. (2020). Driving factors of land change in China's loess plateau: Quantification using geographically weighted regression and management implications. *Remote Sensing, 12*. https://doi.org/10.3390/rs12030453

Saha, & Kukal. (2013). Soil structural stability and water retention characteristics under different land uses of degraded lower Himalayas of North-West India. *Land Degradation and Development, 2013*. https://doi.org/10.1002/ldr.2204

Silva, M. A., & Torres, J. (2022). Intersection between environmental data science and the R community in Latin America. *Environmetrics, 33*(1), e2731. https://doi.org/10.1002/env.2731

Suding, K. N., & Hobbs, R. J. (2009). Threshold models in restoration and conservation: A developing framework. *Trends in Ecology & Evolution, 24*(5), 271–279. https://doi.org/10.1016/j.tree.2008.11.012

Tesfahunegn, G. B. (2013). Soil quality indicators response to land use and soil Management Systems in Northern Ethiopia's catchment. *Land Degradation and Development, 27*. https://doi.org/10.1002/ldr.2245

Welemariam, et al. (2021). *Effect of land use change on soil carbon stock and selected soil properties in Gobu Sayyo, Western Ethiopia*. https://doi.org/10.21203/rs.3.rs-586121/v1

Willy, et al. (2019). The effect of land use change on soil fertility parameters in densely populated areas of Kenya. *Geoderma, 343*. https://doi.org/10.1016/j.geoderma.2019.02.033

Chapter 10
Air and Water Quality Improvement

Abstract The minimal important difference (MID) plays an important role in improving the quality of air and water by setting a limit on the smallest quantifiable change in the environment that would be beneficial for health and ecology. Although air quality indices (AQIs) serve as an important tool for disseminating information to the public and modifying their behavior, these systems have critical limitations. Their categorical structure and static thresholds do not appropriately account for and address numerous emerging risky health impacts of heightened exposure to very low concentrations of pollutants, particularly among children and individuals suffering from various forms of chronic illness. Certain standard values imposed do not take into consideration the state of social welfare and geographic diversity in vulnerability and adaptability to pollution.

MIDs could improve the regulation of air pollution by linking interventions to tangible health improvement outcomes and even more so when using metrics like disability-adjusted life years (DALYs). While useful at the population level, DALYs greatly underappreciate the individual's complexity in health, culturally relevant differences, nonclinical contextual factors like social networks, peer support, mental resilience, and well-being.

The application of MIDs with DALYs in multivariate, personalized frameworks would integrate individual clinical outcomes with systemic public health interventions, resulting in more precise and impactful policies in the context of environmental health.

One such example is the historical Ganga Pollution Case in India, where legal action led to an environmental policy change by the control of a specific source of pollution-industrial effluents. Had defined MIDs been established, they could have served as objective thresholds for health or ecological damage, allowing both sides to scientifically argue within those bounds. MIDs can still provide objective yet quantifiable frameworks for supporting or disputing claims and, as such, are critical in contemporary litigation around environmental issues.

The same concepts can be applied to the management of water quality. In ecological eutrophication and biotic attrition caused by biodiversity loss, nitrogen and phosphorus are recognized as major pollutants which result in ecosystem

I. Fekete, *Transforming Data into Informed Decisions across Clinical and Non-Clinical Domains*, https://doi.org/10.1007/978-3-032-02595-1_10

degradation. Setting nutrient MIDs allows policymakers to define realistic and scientifically sound water quality improvement goals.

Integrating MID with air and water quality standards, policymakers can avoid excessive or ineffective regulation and interventions that are critical advancements to health improvement, ecology, and society.

Keywords Minimal important difference · Air quality · Urban water quality · Pollution control · Air quality indices (AQIs) · Environmental regulations · MID in environmental monitoring · Economic impact · Environmental regulations

The Role of MID in Air and Water Quality Improvement

At the beginning of the book, we have already touched upon air quality index (AQI) and noted that a statistically significant decrease in AQI might not be perceived as a noticeable decrease by patients or doctors (i.e., a noticeable decrease in hospitalizations reported by the MDs) unless it reaches the established minimal important difference (MID). I used this example to illustrate the need for MID. MID can also be utilized as a measure in assessing air and water quality improvements, as it determines the smallest change in pollutant levels that is meaningful/harmful to public health and environmental quality (de Hollander et al., 1999). de Hollander et al. (1999), for example, dealt with the combined effect of air pollution, residential noise, and large technological risks that have hazardous environmental exposure to human health.

The study by de Hollander et al. (1999) proposes an aggregate "public health indicator," a single measure, which would represent the summative effect of multiple negative environmental exposures on public health such as those mentioned before. Such a single indicator might be based on metrics like morbidity, mortality, or other PROs and quality of life (QoL) indicators. This singular indicator can be conceived of as an MID-like in my opinion, back in 1999.

In my mind, a lot of factors such as patient baseline characteristics, region (e.g., highly polluted areas such as certain cities in India or the Alps in Switzerland), or even culture are not included in this single indicator, being a point of criticism of this measure. Since 1999, MID/MCID has refined a lot.

de Hollander et al. (1999) employ a calculation based on the idea of disability-adjusted life years (DALYs) for short, to calculate a threshold analogous to MID/MCID. Weighted by the severity of health impairments, DALYs are computed by adding up the time lost (i.e., total number of years) as a result of early mortality, illness, and time lived with disability.

For every health consequence associated with an environmental exposure, this procedure considers three primary factors: *population affected* (N) is the number of individuals in a population, who are in danger of suffering negative health impacts due to exposure to a particular environmental hazard, such as air pollution or contaminated water. Next, on a scale from 0 (perfect health) to 1 (death), the *severity weight* (*S*) measures the degree of health deterioration. It considers the impact of a health condition on QoL by discounting the amount of time spent with it.

For example, moderate respiratory irritation may have a low weight (e.g., 0.1), whereas severe chronic bronchitis, which is more debilitating, would have a larger weight (e.g., 0.5). Data on environmental exposure levels and epidemiological studies on the relative risk of health consequences linked to that exposure level are combined to estimate the impacted population.

The next health outcome is the *duration of the ailment* (*D*), which is calculated as the average amount of time a person has the ailment or the decrease in life expectancy brought on by early mortality from exposure. This frequently entails comparing the deceased's age with the average life expectancy in order to determine the number of years of life lost in tragic outcomes. It entails figuring out how long a person experiences diminished health as a result of a handicap under nonfatal circumstances.

Third, on a scale from 0 (perfect health) to 1 (death), the *severity weight* (*S*) measures the degree of health deterioration. It considers the impact of a health condition on quality of life by discounting the amount of time spent with it. For example, moderate respiratory irritation may have a low weight (e.g., 0.1), whereas severe chronic bronchitis, which is more debilitating, would have a larger weight (e.g., 0.5). Their formula can be stated as follows: DALY exposure is equal to DALY exposure × DALY exposure = $N \times D \times S$.

This formula calculates the total DALYs lost annually for each exposure by multiplying the number of affected individuals (*N*) by the duration (*D*) and severity weight (*S*). The total health burden, which is the result of adding up these DALYs from various exposures, is used as a cutoff point or standard for evaluating and ranking environmental health interventions.

In line with the idea of MID, which holds that even minor changes in pollutant levels can have a major influence on QoL, and that this threshold aids policymakers in identifying the smallest health impacts that are clinically and socially relevant. In my opinion, beside patient characteristics, DALY should be incorporated in multivariate models when computing MCID for a clinical indication. In addition, time-series analyses should be conducted because the whole ecosystem is undergoing change, and these models should keep track with these changes.

The DALY statistic, although extensively utilized for evaluating the global illness burden and prioritizing health initiatives, has encountered much criticism from multiple viewpoints. In the first place, DALY reduces mortality and morbidity to a singular value, which might mask complex multidimensional health issues, and assumes that all facets of disability and QoL are quantifiable on one continuum (i.e., that singular value). In my opinion, this approach ignores the particularities of the person's context in every sense of the word (clinical context, region, comorbidities, SES, etc.).

Disability weights often come from surveys and expert panels, which is already problematic. These weights may not capture the different sociocultural, ethnic, and societal notions of disability, thus making DALYs more reliable but much less valid.

Additionally, mental health, coping mechanisms, and peer support could also be incorporated into this "MID," if I may call it that.

Lastly, DALY neglects the socioeconomic, environmental, and cultural context of PROs, as already mentioned. To exemplify, the consequences of a handicap may be greater in a resource-poor environment than in a resource-rich environment due to the differences in available care and adequate social support.

Fourth, DALY, as a population-level metric, frequently does not correspond with the individual-level emphasis of personalized or precision medicine. Therefore, my idea is to include it in a multivariate model that is personalized because DALY disregards changes in patient preferences, experiences, patient baseline characteristics or QoL, factors that are essential for individualized care. My proposal is to incorporate patient characteristics (e.g., gender, age, baseline score on the clinical indication, socioeconomic status, comorbidities, region, and cultural background.

Integrating DALY may facilitate the connection between clinical outcomes (assessed using MCID) and their wider societal implications, providing a more comprehensive knowledge of treatment advantages. Modifying MCID calculations with DALY-derived weights may guarantee that treatments emphasize both clinical importance and the reduction of wider socioeconomic and health disparities.

Regarding air quality, MID helps establish the level of particulate matter and nitrogen dioxide emissions that, when reduced, would improve health conditions, thereby assisting policymakers in urban areas (Dockery et al., 1993). Dockery et al. (1993) have also highlighted that when the association between air quality and mortality was established, smoking and other health conditions were not controlled. I am deliberately citing these examples from the 1990s just to show how the metric can be improved and what mistakes should be avoided.

In sum, MID assists in water quality management by facilitating the establishment of realistic contaminant reduction targets necessary for the protection of biological water quality and human health safety.

Economic Impact and Implications of Air and Water Quality Regulations

Water quality is an issue also, especially in India. To illustrate, the so-called Ganga Pollution Case (MC Mehta vs Union of India, 1986[1]) was a lawsuit when environmental activist attorney M.C. Mehta submitted a Public Interest Litigation (abbreviated as PIL). He confronted the significant contamination of the Ganga River due to untreated industrial effluents and urban waste (Supreme Court of India, M.C. Mehta and Anr vs. Union of India & Ors on December 20, 1986). Manufacturing facilities and tanneries were discharging toxic chemicals into the Ganga. This practice

[1] https://indiankanoon.org/doc/1486949/

presents significant (i) health hazards to millions, and (ii) also jeopardizing the ecological equilibrium.

The Supreme Court of India mandated the shutdown of industrial operations lacking adequate treatment systems and instructed towns to establish sewage treatment plants. The PIL led to observable enhancements in certain domains, including a decrease in biological oxygen demand (BOD) levels and stricter industrial controls. The lawsuit established a precedent in jurisprudence for environmental justice in India, underscoring the significance of court involvement in public health and ecological conditions.

I firmly believe that in such cases MIDs as objective measures can be submitted even in civil or even criminal court trials as evidence. Think of sensors that measure pollutants accumulated in the river. Both parties in the courtroom can argue for or against their standpoint once it can be shown that the measured values are below or above the prespecified MID. But, again, I have to stress that this type of evidence was not available in the 1980s.

In my mind, this case illustrates how litigations can gain from MIDs. Experts could and should be involved in the litigation to measure the threshold of damage and harm to health and the ecosystem and incorporate the MID in their argumentation. Advocates might even argue relying on expert opinions (who might have access to already existing MIDs) that the factories should not be demolished, only the amount of toxic material.

MID, when applied to environmental policies, can and would assist in identifying thresholds at which incremental enhancements in air or water quality produce substantial health, ecological, or societal advantages. This strategy allows policymakers to prioritize initiatives that are both successful and economically efficient.

Burnett et al. (2014) suggest thresholds for particulate matter (PM2.5) exposure in air quality management that are associated with quantifiable health benefits. However, reductions in PM2.5 concentrations have been linked to small gains in life expectancy (Pope III et al., 2009). The study by Burnett et al. (2014) constructs a risk function to estimate the global health burden associated with exposure to PM2.5, outlining critical limits of concern for ambient air pollution and its health impact relationship.

Pope III et al. (2009) illustrates the extent to which life expectancy can dramatically increase from even the smallest reductions of particulate matter in the air. This "evidence" illustrates the significance of improving air quality. "An increase of 10 $\mu g/m^3$ in the PM2.5[2] concentration was associated with a reduction in life expectancy of 0.46 ± 0.22 years ($p = 0.039$) from 1978 to 1982 and 0.37 ± 0.20 year ($p = 0.091$) during 1997 to 2001."

What stands out most to me is the focus on air pollution as the only variable controlling life expectancy changes. In other words, my criticism is that the reduction is minute and factors other than changes in air pollution must have influenced the changes in life expectancy (which were not mentioned or controlled). In my

[2] PM = Particulate Matter, 2.5 = Diameter in micrometres (μm).

opinion, it cannot even be ruled out (for the second result) that the reduction in life expectancy is a statistical fluctuation (i.e., statistical noise). I must also mention that their finding ("0.37 ± 0.20 year ($p = 0.091$)" is not even significant, which should have been emphasized in their article. I should also think that it is highly unlikely that their other finding ("0.46 ± 0.22 year ($P = 0.039$)") reaches the MID value. I did not take the effort to see the average life expectancy in the 1970s and 1980s but other factors such as diseases could have also contributed to this seemingly relevant reduction (which were untreatable at that time in India).

Moreover, let us delve further into AQI. I would like to bring another argument. The US EPA (Environmental Protection Agency, 2021, 2022) and numerous international organizations utilize the PM2.5 concentration to evaluate the AQI. As you can see in Table 10.1, the transition from one category to another is 50 in terms of AQI. Pope III et al. (2009) report that "an increase of 10 μg/m^3 in the PM2.5 concentration was associated with a reduction in life expectancy."

The problem, in my judgment, is that an increase of 10 μg/m^3 in the PM2.5 concentration remains in the same health category, not leading to a substantial health improvement. The relationship between PM2.5 concentration and AQI is given as follows:

The US environmental protection agency (EPA) manages the PM2.5 AQI Monitoring System using a well-established framework intended to inform the public and in order to protect health. However, its underlying methodologies have come under increasing scrutiny, particularly from the perspectives of public health ethics, environmental justice, and scientific validity. While the AQI simplifies complex air quality data into accessible categories, several systemic and conceptual criticisms challenge its current adequacy to evolving environmental and public health realities (based on personal communication with government officers whom I would not like to name).

One significant concern is the lack of up-to-date scientific policy evidence used in determining the AQI's PM2.5 breakpoints. Many scholars argue that the epidemiological foundations upon which the breakpoints rest are outdated and misaligned with current evidence on population-level health risks. Notably, the designation of 12 μg/m^3 as the annual mean limit for PM2.5—used to classify "Good" air quality—has been contested, especially considering newer studies demonstrating adverse health effects even at lower concentrations (Di et al., 2017). This raises

Table 10.1 Air quality index (AQI) based on PM2.5 concentrations in micrograms per cubic meter (μg/m^3), following US EPA guidelines

PM2.5 (μg/m^3)	AQI range	Air quality	Health effects
0–12	0–50	Good	No risk
12.1–35.4	51–100	Moderate	Minor risk for sensitive groups
35.5–55.4	101–150	Unhealthy for sensitive groups	Respiratory symptoms
55.5–150.4	151–200	Unhealthy	Increased risk for everyone
150.5–250.4	201–300	Very unhealthy	Health alert
>250.5	301+	Hazardous	Emergency conditions

concerns, particularly for sensitive populations such as children, the elderly, and individuals with chronic illnesses, who may be disproportionately affected by these outdated thresholds.

Second, empirical evidence supports a non-threshold model, suggesting that even minimal PM2.5 exposure can cause harm (Burnett et al., 2014). Consequently, categorizing air as "Good" (0–12 $\mu g/m^3$) under the AQI may falsely convey a sense of safety, thereby underestimating both real and perceived health risks. There is also no justification for the observation period or whether the location of the sensor might have had an effect.

Third, the temporal averaging methods employed in the AQI also present a major limitation. The system typically reports PM2.5 concentrations based on 24-h and annual averages, which can obscure short-term spikes in pollution that pose acute risks (not just for the vulnerable population). For individuals with asthma, COPD, or cardiovascular conditions, brief but intense exposure to PM2.5 can be life-threatening. However, such transient peaks are often masked by averaging procedures, resulting in an underestimation of immediate health threats. In my opinion, there should be an inbuilt indicator that "sounds alarm" when there is a spike in AQI. Such spikes could be identified by employing a predefined MID, in my view.

Another issue can be the failure to account for the AQI's use of uniform benchmarks across populations, neglecting variability in baseline health conditions and environmental vulnerabilities (age, diseases, etc.). Communities residing near industrial zones or in economically disadvantaged urban areas may face compounded exposure risks due to a combination of environmental, socioeconomic, and health-related factors. Yet, the AQI remains largely blind to these context-specific realities (most probably because the measuring sensor is far off from these areas), a topic which I have already touched upon (the location of the sensor). The most affected populations are neglected disproportionately (children and individuals with comorbidities).

Furthermore, the oversimplification inherent in the AQI's structure poses another critical issue. While the color-coded and interval-based system is useful for public communication for its fancy design, it reduces a continuum of risk into discrete, seemingly rigid categories. For example, the transition from an AQI value of 99–101 appears to signify a meaningful escalation in health risk just because we "jump" from one category to another.

However, in reality, this numerical difference may not represent a clinically significant change, as I have already touched upon this issue. Thus, the discrete thresholds fail to reflect the gradient nature of risk and exposure over time.

In conclusion, while the EPA's AQI system has served as a valuable tool for raising public awareness and guiding short-term behavioral adjustments, it increasingly falls short in safeguarding vulnerable populations. The criticisms outlined above underscore the need for a comprehensive reassessment of the AQI framework and the points I have raised.

I would like to return to the odd–even car policy intervention that I mentioned as an example at the beginning of the book. Kumar et al. (2017) illustrate that the odd–even car policy in Delhi briefly lowered air pollution levels, indicating an intervention that may achieve the MCID for air quality. However, Nagpure et al. (2014), consistent with my opinion, warn that in the absence of enduring policies, these modifications do not yield long-term economic or health advantages, highlighting the necessity for cohesive, cost-efficient solutions. Such interventions would also be in the interest of stakeholders, politicians, and the whole healthcare system.

The perspective of integrating MID into air quality regulatory decisions may improve cost-dominated framework. According to Pope III et al. (2009), areas within the United States that experienced a drop in PM2.5 concentration spent 15% more than life expectancy. This 15% value reminds me of the IQWiG threshold (IQWiG, 2023).

Although these data indicate statistically substantial advantages, I would like to highlight the possibility of confounding variables. Utilizing MID as a benchmark enables regulators to more effectively evaluate if changes in air quality result in clinically significant health outcomes, hence preventing overregulation that imposes excessive economic burdens without considerable health benefits. We have to identify the root of the problem and decrease the pollution there as in the Ganga case.

It would also follow that the same form of methodology proposed in MCID/MID could be incorporated in developing water quality laws. Carpenter et al. (1998), along with Dodds and Smith (2016), have focused on the thresholds of nutrient pollution, mainly nitrogen and phosphorus, which when exceeded cause eutrophication and loss of biodiversity to start and have an ecological focus. To put MCID in ecological terms, it would be determining the minimal nutrient concentration reduction that would yield positive changes in water quality and ecosystem health. This ensures that the solutions proposed will have environmental and economic value. The EPA (2007) employs the stressor–response model to set MCID-like nutritional limits, as well and use estimates of the range over which measurable action is expected for remorsefully regulated boundaries.

All of the regulations set are not entirely efficient due to the changes in perception from the general public. de França Doria (2010) emphasizes, in order to have people's support for actions, changes in water quality must be appreciable. MID can also be designed to suit environmental situations so that regulatory measures taken do lead to significant changes that meet public expectations. Phrase "propose nutrient limits for coastal ecosystems in order to mitigate eutrophication," used by Howarth and Marino (2006), suggest incorporating nutrients into "scientific MCID/MID policies," references that suggest that thresholds would be the policy goal and science needs to implement MCID/MID.

Focusing on air and water pollution laws in relation to MCID/MID allows policymakers to focus on initiatives, which provide real benefits (maybe making use of the proposed MID), thus reducing the chances of overregulation while at the same time improving social, environmental, and economic effects. This approach combines cost-sensitive and public interest components of policy by ensuring extensive and measurable impacts on public health and environmental policy.

Strategies for Improving Water Quality in Urban Areas: The Role of MID

Industrial discharges, stormwater pollution, and urban sprawl tend to degrade water quality in major cities (metropolitan areas). Understanding and utilizing the concept of MID could greatly enhance the way water quality improvement projects are designed and evaluated to address these challenges.

In this context, MID can be defined as the lowest detectable change, MDC (minimal detectable change), or the smallest quantifiable enhancement in water quality measures (e.g., nutrient concentrations) that results in significant ecological or public health advantages.

The limits of phosphorus and nitrogen that stimulate noteworthy ecological responses like water eutrophication, algal blooms, and biodiversity loss were identified by Carpenter et al. (1998). In MID, the attribution of values to water quality involves identifying the nutrient concentration thresholds that cause systems to move from one health state to another, for instance, from oligotrophic to eutrophic. These limits are useful to set policy bans and control the level of pollution (Carpenter et al., 1998).

de França Doria (2010) examined the correlation between public perception of drinking water quality and variations in water conditions. An MID for urban water quality may signify the least visible enhancement in the taste, clarity, or safety of drinking water that the public considers significant, offering insights into consumer priorities and promoting public trust in interventions. Public trust can in turn be measured on a Likert scale or Visual Analog Scale (VAS) to see if the intervention has been successful or not.

Stream eutrophication is caused by the over-enrichment of nitrogen and phosphorus, and these nutrient levels were classified as nutrient thresholds by Dodds and Smith (2016). As biological, ecological, and ecosystem MIDs, these thresholds assist urban water managers formulate constructive nutrient reduction targets that improve the health of the water ecosystem.

These thresholds can also be used to design green infrastructure projects, like constructed wetlands and permeable pavements, for urban stormwater management programs that aim to reduce nutrient loads.

Subtle changes in the way headwaters are managed, such as revegetating bare soils and establishing buffer strips, have been shown by Dodds and Oakes (2008) to significantly improve the quality of water downstream. These interventions correspond with the MID concept by recognizing minor but significant activities for watershed health.

Dodds et al. (1998) introduced a classification system for stream trophic states predicated on amounts of chlorophyll, nitrogen, and phosphorus. These thresholds are comparable to MIDs as they define transitions between levels of stream productivity, assisting urban planners in regulating nutrient inputs.

Organizations such as the US Environmental Protection Agency (EPA) offer protocols for establishing nutritional standards grounded in stressor–response dynamics and reference circumstances (EPA, 2007). Establishing MIDs for essential contaminants can assist urban authorities in enacting scientifically grounded water quality requirements.

The assessment of certain undertakings, like nutrient trading schemes and industrial discharge limits, is MID-focused as it analyses how such initiatives improve the quality of water and the health of the ecosystem. To me, MID-based boundaries serve as "criteria" within a monitoring program implementation and provide certainty that interventions will meet ecological outcomes to enable management actions, in case the results are unsatisfactory, to be undertaken.

This idea of MID integrated with public perception and policy enables the setting of limits as well as evaluation of interventions for urban water quality systems. Public perception (de França Doria, 2010) combined with ecological thresholds (Carpenter et al., 1998; Dodds & Smith, 2016) and regulatory easements (EPA, 2007) allows urban water managers to formulate clear and practicable plans to enhance water quality and maintain the ecosystems that reside in the water.

References

Burnett, R. T., Pope, C. A., Ezzati, M., Olives, C., Lim, S. S., Mehta, S., et al. (2014). An integrated risk function for estimating the global burden of disease attributable to ambient fine particulate matter exposure. *Environmental Health Perspectives, 122*(4), 397–403. https://doi.org/10.1289/ehp.1307049

Carpenter, S. R., Caraco, N. F., Correll, D. L., Howarth, R. W., Sharpley, A. N., & Smith, V. H. (1998). Nonpoint pollution of surface waters with phosphorus and nitrogen. *Ecological Applications, 8*(3), 559–568. https://doi.org/10.1890/1051-0761(1998)008[0559:NPOSWW]2.0.CO;2

De França Doria, M. (2010). Factors influencing public perception of drinking water quality. *Water Policy, 12*(1), 1–19. https://doi.org/10.2166/wp.2009.051

De Hollander, A. E., Melse, J. M., Lebret, E., & Kramers, P. G. (1999). An aggregate public health indicator to represent the impact of multiple environmental exposures. *Epidemiology (Cambridge, Mass.), 10*(5), 606–617. https://doi.org/10.1097/00001648-199909000-00030

Di, Q., Dai, L., Wang, Y., Zanobetti, A., Choirat, C., Schwartz, J. D., & Dominici, F. (2017). Association of short-term exposure to air pollution with mortality in older adults. *JAMA, 318*(24), 2446–2456. https://doi.org/10.1001/jama.2017.17923

Dockery, D. W., Pope, C. A., 3rd, Xu, X., Spengler, J. D., Ware, J. H., Fay, M. E., Ferris, B. G., Jr., & Speizer, F. E. (1993). An association between air pollution and mortality in six U.S. cities. *The New England Journal of Medicine, 329*(24), 1753–1759. https://doi.org/10.1056/NEJM199312093292401

Dodds, W. K., & Oakes, R. M. (2008). Headwater influences on downstream water quality. *Environmental Management, 41*(3), 367–377. https://doi.org/10.1007/s00267-007-9033-y

Dodds, W. K., & Smith, V. H. (2016). Nitrogen, phosphorus, and eutrophication in streams. *Inland Waters, 6*(2), 155–164. https://doi.org/10.5268/IW-6.2.909

Dodds, W. K., Jones, J. R., & Welch, E. B. (1998). Suggested classification of stream trophic state: Distributions of temperate stream types by chlorophyll, total nitrogen, and phosphorus. *Water Research, 32*(5), 1455–1462. https://doi.org/10.1016/S0043-1354(97)00370-9

Howarth, R. W., & Marino, R. (2006). Nitrogen as the limiting nutrient for eutrophication in coastal marine ecosystems: Evolving views over three decades. *Limnology and Oceanography, 51*(1_part_2), 364–376. https://doi.org/10.4319/lo.2006.51.1_part_2.0364

IQWiG. (2023). *General methods version 7.0.* https://www.iqwig.de/methoden/general-methods_version-7-0.pdf

Kumar, P., Gulia, S., Harrison, R., & Khare, M. (2017). The influence of odd-even car trial on fine and coarse particles in Delhi. *Environmental Pollution, 20*, 20–30. https://doi.org/10.1016/j.envpol.2017.03.017

Nagpure, A. S., Gurjar, B. R., & Martel, J. C. (2014). Human health risks in national capital territory of Delhi due to air pollution. *Atmospheric Pollution Research, 5*(3), 371–380. https://doi.org/10.5094/APR.2014.043

Pope, C. A., III, Ezzati, M., & Dockery, D. W. (2009). Fine-particulate air pollution and life expectancy in the United States. *New England Journal of Medicine, 360*(4), 376–386. https://doi.org/10.1056/NEJMsa0805646

U.S. Environmental Protection Agency. (2021). *National ambient air quality standards (NAAQS) for particulate matter (PM).* Last modified 2021. https://www.epa.gov/pm-pollution

U.S. Environmental Protection Agency. (2022) *Air quality index (AQI) basics.* Last modified 2022. https://www.airnow.gov/aqi/aqi-basics/

U.S. Environmental Protection Agency (EPA). (2007). *Nutrient criteria technical guidance manual: Rivers and streams.* Retrieved from https://www.epa.gov/sites/default/files/2018-10/documents/nutrient-criteria-manual-rivers-streams.pdf

Chapter 11
MID in Sensor Data

Abstract The concept of the minimal important difference (MID) is vital in evaluating sensor data in many domains, well beyond its roots in clinical research. MID is described as the baseline change in a parameter that can be regarded as useful, meaningful, relevant, or measurable. This criterion distinguishes between nominal changes—usually due to noise or *p*-inflation—and meaningful changes that require action.

In sensor applications such as environmental monitoring, security, agriculture, and health care, the concept of MID offers practical signal interpretation. For example, temperature sensors in greenhouses or fire detectors need to identify meaningful changes amid random fluctuations. In security, a motion detector must distinguish human activity from minor disturbances caused by pets or wind. Therefore, the calibration of MID enhances operational efficiency by reducing false positives.

Sensors in biology show how MID has significant effects in one more area. These sensors incorporate biological elements and transduce the associated activity into digital signals. Other examples include glucose monitors for diabetes and heart rate monitors in telemedicine. As biological systems exhibit variability, the MID cannot be arbitrary in these devices—too much or too little can be problematic for signaling and patient safety would be at stake. As such, tailored calibration alongside patient-reported adjustment is essential for these systems in fine-tuning the limits of deficiency and excess toward optimization.

Health technologies and wearables apply MIDs with the same accuracy to interpret data for patients and clinicians alike. Therefore, wearables gather data on steps, sleep, and heart rates, to name a few. These metrics bear varying degrees of importance to the user depending on clinical context. Baseline health, age, objectives, gender, and many other personal traits together shape what a defined meaningful difference would flag as a substantial change that requires action or immediate medical attention.

Within the scope of air and water quality monitoring, MID specifies the threshold at which a pollutant change necessitates intervention for public health measures to take action. Sociocultural factors like social responsibility also play a role in how conservative or liberal MIDs are set. For instance, Sweden tends to adopt lower MIDs as a result of prioritizing environmental issues, while other regions, for

I. Fekete, *Transforming Data into Informed Decisions across Clinical and Non-Clinical Domains*, https://doi.org/10.1007/978-3-032-02595-1_11

economic reasons, might tolerate higher thresholds. This discrepancy undermines the notion of a singular standardized MID applicable to all sensors.

Telemedicine further illustrates the value of MID in real-world applications. Monitoring patients with chronic illnesses requires the differentiation of harmful changes from genuine indicators of risk. HeartLogic, for instance, anticipates heart failure events by employing multi-sensor data to trigger preemptive (or preventive) measures. Many of these systems, however, define no MID value at all which endangers patients or increases the risk of overwhelming the health-care system with unnecessary alerts. With accurate adjustment to a patient's baseline characteristics, these false alarms can be avoided. To summarize, MID is the focal point enhancing the flexibility and context adaptability of data coming from sensors, wearables, biosensors, and digital health technologies.

Keywords MID in sensor data · Biosensors · Thresholds and MID · Nanostructured materials · Digital health technologies · MID in wearables · MID in telemedicine · Clinical diagnostics

Sensor Data and MID

In the case of sensor data, the notion of MID is a crucial aspect in the evaluation and interpretation of data collected from sensors, with MID having the same definition as in the clinical field (i.e., the smallest change that indicates a meaningful change, e.g., Mouelhi et al., 2020). Sensors are devices that detect and respond to stimuli or other inputs from the physical environment. These inputs and stimuli (e.g., a radiological image, image of the sky to detect fire), or changes, are then converted digitally for further processing. These sensors are related to, but not limited to, temperature, motion (e.g., if a robber is in your house), light and sound (smoke, thunderstorm), or even agriculture (for agriculture, see, e.g., Jiang et al., 2016) among others.

Jiang et al. (2016) in the field of agriculture, including precision farming, monitoring cattle in extensive grazing, tracking commodity movement across the supply chain in real time, automated inventory management, dynamic pricing, and transaction marking mechanisms. Novel data streams generated from so-called global agriculture sensor networks (GASN) face a plethora of challenges, including redundancy, reliability, synchronization, and incorporation, as per Jiang et al. (2016).

Understanding the MID in this context can be challenging, but it is of vital importance given the examples mentioned earlier. Consider fire sensors (that detect smoke), wearables, heat sensors (to locate human beings in darkness), etc.

Sensors have to differentiate important changes from natural "noise," such as normal wind, for example. Sensor data is often "contaminated" with insignificant changes that may be a result of minor environmental changes or simple "system noise." Being able to identify and separate these insignificant changes from the ones that matter is the key challenge in determining the MID. An anchor-based approach might be helpful (e.g., Mouelhi et al., 2020).

In a practical scenario, consider a temperature sensor setup in a greenhouse. A minor fluctuation in temperature—though statistically significant—may not be practically relevant, but a change of a few more degrees could mean the difference between a healthy crop and a ruined one. In this case, the MID would be the *smallest change in temperature that would significantly impact the crop's health.* The reader should now realize that the concept is basically the same for sensors as it is for the MCID. I just call the threshold as MID because many sensors are not used in the clinical field, except for biosensors.

Another example could be a motion sensor installed for security purposes in a flat. This sensor can, of course, make false alarms (Type I error), for example, if a mouse sneaks into the garage (an overreaction) as opposed to a robber (which is a genuine signal). The MID would be the smallest movement that indicates a potential security threat. The MID threshold would separate the false alarm (which is below the MID threshold) from benign movements such as a curtain fluttering in the wind or a pet moving around. Accuracy in these sensors is crucial because false alarms can cost a lot of money.

The so-called environmental sensing has grown recently due to the escalating urgency of climate change and the need for precise monitoring of environmental variables. There are sensors to monitor AQI, for example. In this context, the MID corresponds to the smallest change in AQI that is considered significant for a clinical endpoint, such as the frequency of respiratory diseases over a specified span of time or the number of hospitalizations over a specific time window. An important aspect can be where the sensor is installed in the city because this might greatly influence the values.

Let us consider another example: a sensor monitoring water quality in a river. *If the MID is set too high, the sensor might fail to detect minor but potentially harmful changes in pollutant levels.* This could lead to a situation in which necessary measures (e.g., corrections, modification, reactions) are missed. If we follow this thought further, then this failure could lead to disastrous consequences for the ecosystem and, most importantly, for human health. Experts should be consulted regarding the adjustment of such sensors.

A society, for example, Sweden, that places a high value on environmental protection may opt for a lower MID, despite the higher costs and risks of false alarms. A society that prioritizes economic growth (despite the pollution) over environmental issues may be willing to accept a higher MID.

This societal dimension is yet another aspect that makes the concept of MID subjective and hence speaks against standardization. The same sensor might have completely different MIDs. Let us take the example of Sweden. Leading in renewable energy, Sweden uses 50% of renewable sources. The country makes large investments in wind and bioenergies. It is among the leaders globally in recycling. Less than 1% of household waste is landfilled, the rest is either recycled or transformed in waste-to-energy facilities.

Sweden is ranked fourth in the world for its set targets on carbon emissions. The country aspires to become net-zero in greenhouse gas emissions by 2045, making it one of the most ambitious climate targets in the world. For further information on Sweden, consult Naturvårdsverket (2021) and OECD (2022). The same sensor(s) in another country might have other priorities, as already mentioned, so they would calibrate those sensors according to their needs. In my judgment, this speaks against a standardized MID for sensors too.

Biosensors

Biosensors and sensor data have been applied in biological and chemical analysis too (Miecznikowski & Sellers, 2012). Analytical devices that produce an electrical signal in a response to biological activity for the detection and quantification of specific analysis are known as biosensors. They contain two main parts: biological recognition element and a transducer. The biological element, such as an enzyme, antibody, or nucleic acid, binds with the target analyte, while the transducer measures the interaction (Miecznikowski & Sellers, 2012). Depending on what type of biosensor it is, the signal can be *electrical*, *optical*, or *thermal* (Miecznikowski & Sellers, 2012).

Miecznikowski and Sellers (2012) use chemical sensor data analysis along with data pretreatment techniques. Raw data with noise requires data cleansing (background noise elimination). The techniques regarding cleaning sensor data and background noise eliminations are, for example, Wavelet transform techniques that recover from peak hit severity slope change (error); basically, "peak hit detection error" (Miecznikowski & Sellers, 2012).

LDA and PCA are well-known techniques for high-dimensional data analysis. LDA, maximizing intergroup distances, may run the risk of overfitting in small samples but overall has a lower risk of overfitting compared to PCA. In Type I error control in multiple sensor analysis, Type I error rate must be controlled to avoid empirical null hypothesis rejection. Multiple testing is countered by Bonferroni correction, which adjusts significance threshold. False positives (Type I error) are limited by Benjamini and Hochberg's false discovery rate (FDR). Biosensor data require high-level statistical analysis.

Biometric sensors have been part of our life: from facial recognition systems to fitness trackers, all these devices are constantly capturing and processing our biological and biometric information as well as offering feedback (including biofeedback). But what is the practical significance of changes in the data collected?

The primary purpose of biometric sensors is to capture, monitor, and measure biological functions or behaviors like heart rate, blood pressure (BP), or even a fingerprint's minute features. The information collected from these sensors is analyzed and further used in *personalized* decision-making processes, for instance, determining if a person is eligible to exercise or for the identification of a person.

However, biological data is variable: heart rates fluctuate, BP rises and falls, and even fingerprints can change slightly over time, let alone get burnt. People age, they start to wear spectacles, etc. The challenge for biometric sensors is to distinguish between normal variations and significant changes that may indicate a problem or a false reading. This is where the concept of MID becomes critical. If a sensor detects a deviation (i.e., exceeds the MID), it can trigger a false alarm. Such an alarm can occur in a medical center, at the airport, or in the sleeping room of the parents where there is a baby.

Take, for example, a heart rate monitor. If the MID for heart rate is set too high, the monitor might fail to detect potentially dangerous conditions like *tachycardia*. Finding the appropriate equilibrium requires precise engineering, professional medical insight, and ideally, patient interviews, and patient feedback. It entails not just the technical features of the biometric sensor, but also the social dimensions, especially the user's disposition toward excessive false alarms, as well as the risk factors associated with missing an actual alarm.

I would like to stress once more that while false alarms may not be damaging to the patient, their monetary cost can be exceedingly high. Actually, this statement applies to any sensor; think of a fire-alarm sensor that produces a false alarm.

In conclusion, advances in real-time monitoring technologies, such as biosensors and wearable devices (e.g., watches), will enable continuous assessment of biological changes, thereby refining MID estimates. Importantly, I deem *continuous assessment* and *personalization* as highly important, though they might not always be available (e.g., in the clinical field).

Based on their differing methods of transduction, biosensors can be categorized into electrochemical, optical, and mass-sensitive. Optical biosensors employ light for their detection systems, making them useful for certain applications, for instance, in vivo monitoring (Jiang et al., 2016).

The main objective of sensors remains the detection of biomarkers, pathogens, and even pollutants within the environment, giving timely and accurate information that is important for informed and optimal decision-making (Miecznikowski & Sellers, 2012).

Clinical Diagnostics

Using point-of-care testing, biosensors are utilized in clinical settings for immediate testing of patient samples during a clinical diagnosis. Take, for instance, a glucose biosensor, which is readily utilized by diabetic patients for monitoring their blood sugar levels. Such biosensors are based on the principle that glucose oxidase catalyzes the oxidation of glucose. The sensitivity and specificity of these biosensors are critical, as they directly impact patient management and treatment outcomes.

Thresholds and MID

In the case of biosensors, thresholds point out the limits below which an analyte cannot be observed. The importance of these thresholds stems from the role they play in clinical interpretation. The idea of MID is equally important, as it shows how any changes in an observation are achieved.

The thresholds have been set using empirical evidence and clinical practice guidelines. Take, for example, glucose monitoring, where thresholds of 70 mg/dL are often set for diagnosing hypoglycemia, while >180 mg/dL may be suggestive of hyperglycemia. These thresholds aid decisions in treatment approaches and serve as measures to avert diabetes-related complications. Again, thresholds for environmental monitoring of pollution are set in a manner to safeguard public health and to protect the environment from the potential negative effects of exposure to pollution.

A threshold in biosensors represents a minimal criterion for analyzing what occurs or does not occur. For example, a biosensor with specific parameters has the ability to change the classification of a biometric indicator to a more significant one for clinical comparison than the previously accepted standard but if it was those situations where there is no clinical intervention. Different types of biosensors have been developed, and these advances have been actively driven by nanotechnology, materials science, and biochemistry, which have led to the enhancement of biosensor performance.

Nanostructured Materials

The sensitivity and stability of biosensors have been enhanced with the integrated use of nanostructured materials such as carbon nanotubes and graphene. Not only do these materials possess excellent electrical conductivity, but they also have an incredibly high surface area, which increases the transduction efficiency of biosensors. For example, the application of platinum nanoparticles in amperometric biosensors has noticeably improved detection limits for numerous analytes, such as glucose and histamine.

Further innovations in microfluidic technology allow for the design of portable and miniature analytical biosensors that can perform complex analyses using very limited sample volumes. These devices can be particularly useful for point-of-care purposes where rapid results need to be obtained. Moreover, wearable biosensors that track physiological parameters, such as the composition of sweat, are becoming more popular for health monitoring and disease detection.

Moreover, the concept of genetically encoded biosensors integrates biosensing with the use of modified proteins to identify certain metabolites in living cells. These biosensors make it possible to view real-time changes in cellular metabolism and even disease development. The invention of FRET-based biosensors has broadened the scope of activity of genetically encoded biosensors. These sensors are now able to detect multiple analytes at once.

However, as we progress in the technology of biosensors and their applications, we will also face unique hurdles. Questions regarding specificity, sensitivity, and stability will always surround the efficacy of biosensors in clinical and environmental settings.

Specificity and Cross-Reactivity

Development of biosensors comes with the problem of ensuring that cross-reactivity is minimal and that they are specific to a certain analyte. If the biosensor detects compounds that are too similar, there will either be a false positive or a false negative. To address this issue, further inputs are being developed in order to amplify the selectivity of biosensors with the use of new bioreceptors and new materials.

However, with the increasing clinical utility of biosensors, regulatory devices will also be pivotal in their life-cycle realization. There is an emphasis on the establishment of widely accepted methods for the validation and performance assessment of biosensors for their safe and effective application in practice, including employing machine-learning approaches.

Digital Health Technologies

Within digital health, MID is typically viewed as how patients communicate with health tools, how these tools affect their behavior, and, consequently, how that behavior affects the PROs. Mäder et al. (2023), for instance, state that Germany is a prime player in evidence-based digital health and electronic health applications, most probably due to its Digital Health Application (the abbreviation in German is: DiGA) system.

Mäder et al. (2023) set out to investigate the evidence requirements outlined by the Federal Institute for Drug and Medical Devices (BfArM; Bundesinstitut für Arzneimittel- und Medizinprodukte) as the prerequisite for the approval of DiGA and those applications that are permanently listed in the so-called DiGA directory.

It is interesting to point out that in Germany, the authority that is engaged in the early benefit assessment of digital health applications is the BfArM (Federal Institute for Drugs and Medical Devices). As of 2025, BfArM was the largest drug approval authority in Europe. These digital applications, on the other hand, can also be reimbursed by statutory health insurance and prescribed by physicians and psychotherapists in Germany. To relate DiGAs to MCID, BfArM is similar to the G-BA in that it also work with MCIDs. Before submitting a proposal for a new DiGA, there is a hearing where they give advice on the SAP as well as the MCID.

Mäder et al. (2023), using a multipronged methodology strategy, first sought to determine the evidence required for granting a permanent role for DiGA applications, and second, analyzed the available evidence to back these applications up. The results showed that all 13 applications, which were permanently included in the DiGA-directory, were aimed at solving various mental health problems (depression and anxiety), and each application contained proof of the positive effect of its use in the field of health care. Such a situation is currently present (in January 8, 2025), with 68 digital health applications available in the so-called DiGA directory (https://diga.bfarm.de/de/verzeichnis). (As of May 26, 2025, there are 69 digital health applications listed in the database (https://diga.bfarm.de/de/verzeichnis), just for your information).

But how large were these effects? This is why we are now talking about DiGAs. The BfArM also evaluates the medical products such as herbal and homeopathic products but also minerals and vitamins. Notably, all of these applications had conducted RCTs. The manufacturers opted for RCTs, underscoring a commitment to high-quality medical evidence (causality) and indicating the potential of DiGA to significantly improve health care and health. The BfArM can be contacted for an initial counseling session before starting compiling the SAP and the RCT. In this session, experts answer all relevant questions that the manufacturer may have.

In conclusion, the study by Mäder et al. (2023) found that the permanently listed DiGAs not only meet but also exceed the evidence standards required by the BfArM[1] ("Federal Institute for Drugs and Medical Devices" in Germany). This finding highlights the robustness and reliability of Germany's approach to integrating digital health into standard medical care and also aligns with Germany's position as a front-runner in the digital health arena in Europe, setting a high benchmark for evidence-based digital health interventions. Mäder et al. (2023) write that "[] although BfArM accepts study designs with a lower level of evidence for the proof of a positive health-care effect, all manufacturers conducted a study with a high level of evidence."

MCIDs were indication-specific and were reported only in 4 out of 13 DiGAs (see Table 5, Mäder et al., 2023). It is not clear to me why no MCID was reported in the rest of the studies and why BfArM is less strict than the German authority GBA.

BfArM has a list of digital health applications, which can be accessed online (https://diga.bfarm.de/de/verzeichnis). More information on digital health applications can be found at "The Fast-Track Process for Digital Health Applications (DiGA) according to Section 139e SGB V A Guide for Manufacturers, Service Providers and Users" (https://www.bfarm.de/SharedDocs/Downloads/EN/MedicalDevices/DiGA_Guide.pdf?__blob=publicationFile&v=2).

[1] https://www.bfarm.de/EN/BfArM/_node.html

MID in Wearables

Let us review a real-life example in the field of wearables. We are all aware of how these devices have gained popularity recently. They help to track activity levels, sleep hours, pulse, heart rate, and so on. But how does the MID framework come into play here, and what insights can be drawn from it?

These devices allow individuals to quantify, track, and analyze their daily activities. When it comes to wearable fitness trackers, MID can be considered the smallest change in a measured PRO that a user perceives as significant/patient-relevant. This could include changes in the number of steps walked, calories burned, hours slept, and so on. The MID varies from person to person (which makes these devices personalized) and understanding it can help both manufacturers and users alike to interpret the data generated by these devices more effectively. In my opinion, individual differences speak against the use of a standardized metric, as proposed by, for example, the German institute IQWiG (2023).

For instance, let us consider a patient whose goal is to increase his/her daily step count. One individual might regard an increase of 1000 steps as significant, whereas another might only view an increase of 5000 steps as a meaningful change (i.e., his/her MID). These interindividual differences depend on clinical indication, gender, age, etc. (baseline characteristics). However, with the relevant set of information, patients can set achievable goals.

In my opinion, MID is a dynamic concept that needs to be adjusted to individual perceptions, patient characteristics, contexts, and clinical indications. Only then can we truly harness its potential in improving the effectiveness and relevance of health technologies like wearable fitness trackers.

MID in Telemedicine

A particular case that comes to mind is a study that sought to establish the MID for a telemedicine (TM) application used to monitor patients with chronic heart failure (Accorsi et al., 2024). This case (specifically, telecardiology) provides a good example of how the concept of MID can be employed in real-world settings to evaluate the effectiveness of an intervention.

In the study by Accorsi et al. (2024), the TM-application was used to monitor various indicators of the patients' health status. However, no MIDs were used. It is obvious that the challenge lies in determining what constitutes a "significant" clinical change; this is where the concept of MID comes into play: any change less than this would be regarded as noise or natural fluctuation, whereas any change greater than this would be considered a potential warning sign. Without some MID, there is a risk that changes in health parameters could be *misinterpreted*, which can lead to *missed warning signs* or even *overburdening* (the patient gets exhausted) the *heart and the body by performing strenuous activity*. Lastly, *unnecessary interventions* could also be avoided, which in turn would reduce the cost of health care.

TM and wearables point in the direction, in my mind, that the MID is not a fixed value contra IQWiG (2023, p. 173), but rather, it is context-dependent and undergoes much individual variation. It also varies depending on the specific health parameter (i.e., clinical endpoint) being measured, the characteristics of the patient population, and the purpose of monitoring.

Accorsi et al. (2024) cite a study by Boehmer et al. (2017), which uses a multisensor algorithm (HeartLogic, Boston Scientific) that "analyses five metrics-intrathoracic impedance, nocturnal heart rate, presence of a third heart sound, respiration rate, and patient activity-and creates a quantitative and objective assessment of the clinical state."

According to Boehmer et al. (2017), HeartLogic showed a 70% sensitivity score in predicting decompensation events. More importantly, almost 90% of the patients included in the study received an alert nearly 2 weeks before the onset of an actual heart failure event. This is a great achievement in medicine, in my opinion.

MID has been instrumental in bridging the gap between clinical research/practice and patient care. Patients can also give feedback about their health status and participate in SDM (shared decision-making). In health care, MID shifts the focus from individual-neutral medicine to the fine-tuning of treatments based on individual patient needs and responses (i.e., if the patient is a so-called responder or nonresponder), see IQWiG (2023, p. 172). A responder refers to a patient who responds to the therapy (success), the opposite is true for nonresponders as per medical nomenclature (e.g., IQWiG, 2023, p. 172).

References

Accorsi, T. A. D., Santos, G. G. R. D., Nemoto, R. P., Moreira, F. T., De Amicis, K., Köhler, K. F., Cordioli, E., & Pedrotti, C. H. S. (2024). Telemedicine and patients with heart failure: Evidence and unresolved issues. *Einstein (Sao Paulo, Brazil), 22*, eRW0393. https://doi.org/10.31744/einstein_journal/2024RW0393

Boehmer, J. P., Hariharan, R., Devecchi, F. G., Smith, A. L., Molon, G., Capucci, A., et al. (2017). A multisensor algorithm predicts heart failure events in patients with implanted devices: Results from the MultiSENSE study. *JACC Heart Fail, 5*(3), 216–225. https://doi.org/10.1016/j.jchf.2016.12.011

IQWiG. (2023). *General methods version 7.0.* https://www.iqwig.de/methoden/general-methods_version-7-0.pdf

Jiang, et al. (2016). *Application and implementation of private cloud in agriculture sensory data platform.* https://doi.org/10.1007/978-3-319-48354-2_6

Mäder, M., Timpel, P., Schönfelder, T., Militzer-Horstmann, C., Scheibe, S., Heinrich, R., & Häckl, D. (2023). Evidence requirements of permanently listed digital health applications (DiGA) and their implementation in the German DiGA directory: An analysis. *BMC Health Services Research, 23*(1), 369. https://doi.org/10.1186/s12913-023-09287-w

Miecznikowski, J. C., & Sellers, K. F. (2012). Statistical analysis of chemical sensor data. In *Proceedings of advances in chemical sensors* (pp. 327–347). https://doi.org/10.5772/30756

Mouelhi, Y., Jouve, E., Castelli, C., & Gentile, S. (2020). How is the minimal clinically important difference established in health-related quality of life instruments? Review of anchors and methods. *Health and Quality of Life Outcomes, 18*(1), 136. https://doi.org/10.1186/s12955-020-01344-w

Naturvårdsverket. (2021). *Sweden's environmental objectives – An introduction*. Swedish Environmental Protection Agency. https://www.naturvardsverket.se

OECD. (2022). *Environmental performance reviews: Sweden 2022*. Organisation for Economic Co-operation and Development. https://doi.org/10.1787/19900090

Chapter 12
MID in the Social Sciences, Linguistics, Education Sciences, and Sociology

Abstract This chapter offers a fresh application of the minimal important difference (MID) idea in linguistics and the social sciences, among other nonclinical fields, suggesting that it can be applied in methodological frameworks and taught in academic programs, Mostly used in clinical medicine, MID is defined in both challenging and practical terms as the smallest change in a key variable that is relevant and significant. This chapter highlights how social scientists rely excessively on *p*-values and statistical significance, thus producing poor result interpretations and Type I error. They sometimes even ignore an almost zero effect if the result is significant.

Generally speaking, social sciences, including linguistics, can benefit immensely from MIDs, especially in large datasets where small variations may produce statistically significant results despite minute differences. MIDs could simplify the evaluation of change relevance in fields including educational communication, where focused interventions seek clear improvements in learning or communication. Consider a situation in which a particular teaching strategy raises test results by a modest margin; its application would only be justified if the score increase exceeds the MID level, thereby verifying its influence on student performance.

In this instance, this chapter criticizes studies in linguistics and psychology that document rather small outcomes. Large sample sizes and a lack of suitable effect size measurements allow one to treat a difference of 7 ms in reaction time or a 1.19-point shift on a personality scale as statistically significant. Such cases show the possibility of Type I error and the possible data fabrication distortion in the absence of MID.

This chapter investigates how MIDs might improve experimental designs by establishing explicit limits that define meaningful deviation from simple statistical fluctuations in linguistic data, thereby improving the clarity and reliability of linguistic experiments. One suggestion is to specify particular MIDs inside the statistical analysis plan, say half a standard deviation of the mean change as the cutoff.

Moreover, this chapter presents a tailored or stratified method of MID, whereby group-specific thresholds are computed using age, gender, and cognitive status, for example. Particularly in language acquisition, speech disorders, or second language

I. Fekete, *Transforming Data into Informed Decisions across Clinical and Non-Clinical Domains*, https://doi.org/10.1007/978-3-032-02595-1_12

learning research, this kind of method enables complex and customized data analyses.

Furthermore, addressed via the idea of MID are the pragmatic, communicative, and philosophical scopes of language. In both written and spoken forms of communication, even the smallest changes—such as a missing comma or a minor change in intonation—can have significant effects in meaning. This chapter contends overall that the social limits of MID should be extended to increase the scientific rigor, validity, and practical value of social science study. Only then statistics will retrieve its credibility in the human sciences.

Keywords MID · linguistics · Social sciences · Sociology · Psychology · Humanities · Educational sciences · *p*-Hacking · Statistical malpractice · *p*-Inflation · Type I error · False positives · Philosophy of language · Communication sciences

MID Outside the Natural Sciences

This chapter represents one of the novelties of the book—especially in linguistics—because there are no established minimal important difference (MID) metrics in that field. Crucially, MID has not been established in the social sciences either at all. I strongly advocate to include MID in the curricula in those majors at universities and colleges where they pursue quantitative research. Another problem is that not even quantitative research is in the curriculum in many majors, such as mass communication and linguistics. In addition, I am for a personalized MID, which can change over time even within an individual, and which can be modeled by a multivariate statistical analysis.

To the best of my knowledge, MCID-like metrics are rarely or almost never employed in the social sciences or the humanities in curricula. Instead, effect sizes are used (or not even that, unfortunately) to rule out Type I error or to infer the strength of the effect. I have already seen that researchers just report the effect size, which is close to zero, and claim that the effect is small. I regard this as a statistical malpractice.

For example, educational research could benefit a lot from the use of MIDs. For instance, in the *evaluation of the impact of teaching interventions or novel didactic methods on student performance metrics*. To comprehend the essence of MID in the field of social sciences and humanities, imagine a scenario. Thresholds (or cutoffs) could be defined to ascertain what constitutes a meaningful and noticeable improvement in test scores as a result of specific pedagogical strategies or intervention in an experiment or a didactic method. Is it a "real" and noticeable improvement in student performance or "just" statistically significant? The question arises as to whether a statistically significant, yet slight statistically significant improvement is meaningful and impactful to the students, or not (or just statistically significant)? It could be the case that the teacher or one of the relatives as "anchor people" would report if the change is marked and noticeable or not. Also, research in social work could also apply MID-like thresholds to measure the impact of some community-based interventions on QoL indicators.

Linguistics

MID in the realm of linguistics is a completely novel concept, in contrast to the clinical fields. This section explores how MID can further enhance the *quality* and *reliability* of research findings in linguistic studies (e.g., psycholinguistics or applied linguistics), as it would filter trivial (i.e., irrelevant) yet statistically significant differences (even after *p*-adjustment).

Prasad (2024) also emphasizes that there has been an overreliance on *p*-values, leading to the *misinterpretation of clinical findings*. I claim that the same phenomenon has taken place in other disciplines, including linguistics too. Because as I have mentioned earlier, the use of MID in linguistics is an uncharted area, I am going to show one study (Fekete et al., 2018) employing MIDs and two studies as negative examples in which "effects" were found which would not pass the MID thresholds (Bourassa & Besner, 1998; Perea & Lupker, 2003).

In the past, linguistic research used smaller datasets, often collected and annotated by hand (e.g., Brown & Fraser, 1964). However, big data have enabled linguists to process large datasets, which lead to small differences becoming *statistically highly significant* (e.g., Brezina & Meyerhoff, 2014; Lin et al., 2013; Sullivan & Feinn, 2012), a statistical principle ignored even by senior researchers.

A specific example, however, from the realm of linguistics is astounding and alarming. Brezina and Meyerhoff (2014) present evidence demonstrating that sociolinguistic studies based on aggregate data are completely unreliable. Using the British National Corpus (BNC 32), Brezina and Meyerhoff (2014) show that *random* "speaker groupings can often yield statistically significant results." BNC 32 refers to a smaller subset of the BNC, which is a 32-million-word collection of written and spoken English. Koplenig (2019) goes as far as claiming that *statistical significance testing in corpus linguistics should be abandoned*. But, unfortunately, with the advent of AI and the analysis of big data, false positives are inevitably on the rise.

Brezina and Meyerhoff (2014) point out the problem in big data and suggest the Mann-Whitney U test as a solution, yet they do not address MIDs or effect sizes as alternatives, which I believe would add to their arguments and improve the quality of linguistics research findings in general. In my mind, however, I must disagree with Brezina and Meyerhoff (2014) because in big data, the Mann–Whitney U test can also often detect statistically significant differences even when these differences are *practically insignificant*, which is because statistical significance becomes easier to achieve in big data. However, large datasets lead to misinterpretation of findings due to Type I error (e.g., Angst et al., 2017; Brezina & Meyerhoff, 2014; Farivar et al., 2004; Kalinowski & Fidler, 2010; Lin et al., 2013; Sullivan & Feinn, 2012).

The articles in the domain of linguistics which use MIDs are immensely sparse, to date almost nonexistent with a few exceptions (e.g., Fekete et al., 2018). Fekete et al. (2018) use both MIDs and effect sizes to filter out trivial but statistically significant differences. I am going to present their study in more detail in this chapter. But first let us begin with some negative examples which legitimate the use of MIDs.

Crucially, Angst et al. (2017) contend that MCIDs/MIDs reduce bias, which is important for us in the context of linguistics too. In linguistic studies, subjective interpretation bias can occur, for example, when researchers' personal expectations, errors, or cultural backgrounds among others affect the interpretation of linguistic data. Finally, MID thresholds also allow for the *comparability across linguistic studies*, which, in turn, permits comparability in meta-analyses and systematic reviews. This, in turn, enhances the validity of research findings.

When MID is incorporated into linguistic studies (let us call it MID as of now, as this is a novel concept in linguistics), MIDs can help to quantify *nontrivial differences* in language use, language comprehension, corpus linguistics, sentiment analysis, discourse analysis, computational linguistics, psycholinguistics, cognitive linguistics, sociolinguistics, neurolinguistics, historical linguistics, forensic linguistics, second language acquisition, language policy and planning, dialectology, and language development among a bunch of other areas of linguistics. All of these branches of linguistics without exception can benefit from MIDs when quantitative data analysis is involved.

The process of applying MID in linguistic studies begins with the identification of the specific linguistic variable (dependent variable, outcome variable, or response variable) to be measured, just like in the case of clinical studies, analogous to PROMs/PROs or clinical endpoints in their terminology.

This linguistic dependent variable (comparable to PROs) could be anything from vocabulary size of children (number of words known by the child), grammatical accuracy (%) in a psycholinguistic experiment, reaction time to words in milliseconds (RTs; so-called *lexical decision paradigm*; see, e.g., Meyer & Schvaneveldt, 1971), vocabulary frequency, lexical sophistication, pronunciation accuracy (%), accuracy in morphological inflection, speech rate (e.g., number of words per minute), or to the complexity of sentence structure measured on a Likert scale.

I would like to give some other examples for linguistic variables where MID could be useful. For example, phonological variables (whether the participants can distinguish between two forms of pronouncing "r"), accent (whether the participants can distinguish between two accents), or word frequency effects in comprehension.

We have seen that MID is not restricted to any measurement scale; it can be interval or ordinal, for example. Following linguistic data collection, the MID has to be established. But the question is as to how we should determine the MID in a linguistics study. My advice, based on my own experience in HTA, comes from the IQWiG recommendation (IQWiG, 2023) that every single step of the statistical analysis, abbreviated as SAP in HTA, should be *prespecified*. Given this claim, the MID should also be prespecified and not determined in a post hoc manner. Prespecification refers to the practice of setting the steps of statistical analysis in advance before recruiting the participants of the study, just like the prespecification of sample size (sample-size calculation).

MID in linguistics can, for example, (i) involve calculating the average performance or ability level, and then determining the smallest change from this average that would be considered important, relevant, substantial, or noticeable. However, the application of MID in linguistic studies is not established and is not solely a quantitative process as in the case of clinical studies (e.g., Jaeschke et al., 1989; Norman et al., 2003). It also requires a qualitative understanding of language and its usage. What might be considered an MID in one linguistic context may not hold the same significance in another, I should think.

For example, a small change in vocabulary size might have a significant and *noticeable impact* in a second language learning context in adults, but may be less noticeable in a native speaker conversation in a linguistic experiment or in a natural setting. One surefire solution could be adopting half an SD criterion form the clinical field (e.g., Farivar et al., 2004; Norman et al., 2003). In sum, if the difference or change between two conditions or groups exceeds half an SD (or pooled SD) of the baseline scores in an experiment, then we can accept it as a *linguistically relevant* change.

In addition, it is crucial to consider the *practical implications of the MID*. This means considering how the *findings can be applied in real-world situations*. For instance, if the study reveals that an MID in grammatical accuracy (%) leads to a significant improvement in comprehension, this could have implications for language teaching methods. Hence, MID can enhance the practical implications of linguistic research findings by further bolstering the research findings. To put it simply, we will have another valuable statistical tool in our hands along with conventional hypothesis testing.

One study in the field of linguistics that has employed MIDs is by Fekete et al. (2018). Their study utilized MIDs to distinguish between *statistically significant* and *linguistically meaningful/relevant differences* in psycholinguistic data in a cohort of around 400 participants.

The use of MIDs stemmed from the understanding that significant results may often be influenced by factors such as random measurement errors, Type-I and Type-II errors, or sample-size inflation, which can erroneously lead to the rejection of the null hypothesis despite the differences being linguistically irrelevant (e.g., Brezina & Meyerhoff, 2014; Lin et al., 2013; Sullivan & Feinn, 2012). "With a sufficiently large sample, a statistical test will almost always demonstrate a significant difference, unless there is no effect whatsoever, that is, when the effect size is exactly zero; yet very small differences, even if significant, are often meaningless" (Sullivan & Feinn, 2012).

Fekete et al. (2018) employed MIDs to filter out Type I error and trivial differences (Crosby et al., 2003; Mokkink et al., 2010a, b; Revicki et al., 2006, 2008; Engel et al., 2018). In their research, MID was calculated using both a standard deviation criterion and effect size parameters to assess the linguistic relevance/meaningfulness of differences between experimental conditions, such as different linguistic structures in German.

Their study used distribution-based techniques for MID estimation (Crosby et al., 2003). This included the adoption of a one-third standard deviation criterion for MID, based on the recommendations by Farivar et al. (2004) and Eton et al. (2004), although other studies have suggested a one-half standard deviation criterion as a more universally applicable measure (Farivar et al., 2004; Norman et al., 2003, 2004; Watt et al., 2021).

Moreover, as already mentioned, to circumvent the issue of overestimating differences in large samples, the study by Fekete et al. (2018) emphasized the importance of reporting effect sizes. This was done using nonparametric measures like Cliff's delta (Cliff's δ with 95% CI) instead of traditional parametric indices like Cohen's *d*, providing a more robust assessment of the effect magnitude between groups (Cliff, 1993, 1996). In summary, this study applied MID estimates to discern *linguistically meaningful differences*, ensuring that significant findings are both *practically relevant and linguistically meaningful*. This approach enhances the interpretability of results and the quality of the research findings. In conclusion, applying MIDs in linguistic studies is essential to filter out trivial (linguistically not meaningful) differences or changes (Fekete et al., 2018).

Observed estimated differences and adjusted multiple comparisons for the critical contrasts in Experiment 1 in Fekete et al. (2018), originally Table 5 in Fekete et al. (2018). The comparisons are within-participant comparisons. "***" designates $p < 0.001$. The comparisons are *within-participant* comparisons in the German language. Their translation is irrelevant to the subject matter of the book (Table 12.1).

One example where MIDs could have helped to filter out trivial differences is the psycholinguistic study by Perea and Lupker (2003). Surprisingly, the study is not that old. Still, there are serious problems with the methodology employed in the study in my opinion. First of all, in their first experiment, 120 students were recruited arbitrarily without any sample-size calculation, while in their second experiment only 36 students without justification, and in their third experiment, 36 students. More surprisingly, in their fourth experiment only 18 (!) participants were recruited. These arbitrary (!) sample sizes without any sample-size calculation do affect the effects. I would not even accept this sample size at UG level, let alone in a reputable journal.

Perea and Lupker (2003) wish to provide insights into the mechanisms of visual word recognition and the semantic/associative priming effects driven by so-called transposed-letter (TL) nonwords (e.g., "jugde" instead of "judge"). They wanted to find out if TL nonwords primes (e.g., "jugde") would activate semantic/associative information from their base words, such as "judge." However, their study exhibits serious statistical limitations in my opinion. Specifically, the trivial differences, which undermine the reliability and credibility of the research findings.

The small "effects" in their study, particularly with the semantic/associative priming effects of TL nonwords pose a serious methodological problem. The differences in reaction times between conditions are very small. For example, they report semantic/associative priming effects of around 10–15 ms for TL-internal nonwords. Moreover, they claim that "responses to target words were 9 ms faster when these target words were preceded by TL primes than when they were preceded by control

Table 12.1 The application of MIDs in a linguistics experiment employing a within-participant design. The comparisons in the column "Comparison" are within-participant comparisons of various linguistic structures in German. The translations do not matter for understanding the core idea of the table. What is important is that the Fekete et al. (2018) employed MID in conjunction with Cliff's δ (95% CI). The translations are below the German text in the blue boxes.

Comparisons	Observed mean difference	adj. *p*-value (comparison)	MID SD-criteria 1/3 SD (Eton et al., 2004)	Cliff's *δ* (95% CI)	Nature of difference
wer–wer so alles (who–who all)	0.204	9.40E−06 ***	0.3843	0.11 [−0.02; 0.24]	Trivial
wer–die (who–the)	0.366	8.32E−16 ***	0.388	0.18 [0.11; 0.24]	Trivial
wer so alles–die (who all–the)	0.162	2.77E-04 ***	0.3843	0.08 [0.01; 0.14]	Trivial
wer–einige (who–some)	1.05	2.64E−84 ***	0.374	−0.52 [−0.57; −0.46]	*Nontrivial*
einige–die (some–the)	1.416	1.98E−116 ***	0.388	−0.63 [−0.68; −0.58]	*Nontrivial*
einige–wer so alles (some–who all)	1.254	3.63E−101 ***	0.3843	0.59 [0.53; 0.64]	*Nontrivial*
wer so alles–wer alles (who all–who all)	0.282	1.41E−10 ***	0.3977	0.13 [0.06; 0.20]	Trivial

primes" (Perea & Lupker, 2003, p. 835). When it comes to error rates, they write that "the error rate to word targets was 1.1% lower []," again, showcasing the need for MID and effect sizes. The authors seem to ignore any random fluctuation in statistics.

They are also citing a study by Bourassa and Besner (1998) in which a difference of 7 (!) ms was considered significant (Perea & Lupker, 2003, p. 831). It might just be a coincidence that it is the same journal *Memory & Cognition*. Perea and Lupker (2003) themselves acknowledge and confess that "This effect, however, was quite small (7 ms) and required a rather large number of participants in order for it to be significant." They do not even mention how large was the "large number of participants." This statement about a *significant* difference of 7 ms is a methodological manipulation, a typical scientific malpractice, in my mind, and is a classic example of so-called *p*-value hacking, the latter being also known as "data dredging": they basically increase the sample size until the "effect" becomes statistically significant. This is also called "optional stopping" or "data peeking" in the "*p*-hacking literature" (e.g., Stefan & Schönbrodt, 2023). They even go further and write that "The size of our priming effect for the RL-nonword primes when standard nonword foils were used (6 ms) was virtually identical to their 7-ms effect."

Similar findings can be found anywhere, for example, quite recently, unfortunately, by Otero-López et al. (2021). The study focused on the interplay between personality traits, coping strategies, and compulsive buying behavior in 1093 (sic) university students. The study sample was categorized using sociodemographic variables, Big Five personality traits, and coping strategies. Findings revealed that compulsive buying behavior can be predicted by risk factors such as neuroticism and problem avoidance, whereas active coping strategies such as problem-focused coping, cognitive restructuring, and social support lowered the likelihood of it. The research outlines recommendations for the prevention and intervention of compulsive buying in young adults, revealing gaps in deeper inquiry in the discipline.

My first doubt is why Otero-López et al. (2021) took unusually so many participants and why there are no effect sizes and sample-size calculation reported in the article. Second, most importantly, they write that agreeableness was significant in the results section [$t(1091) = 2.47$, df = 1091 $p < 0.014$] (). Otero-López et al. (2021) report a significant difference between non-CB and CB[1] in *International Journal of Environmental Research and Public Health* for the construct Agreeableness (see Table 2): non-CB (mean): 31.24 (SD: 4.27), CB: 30.05 (SD: 4.53).

Moreover, consistent with Table 2 in their article, in Table 1 (Otero-López et al., 2021), it is shown that the non-CB group contains 1007 (sic) participants, whereas the CB group only 86 (sic).

I find it unacceptable to claim that a difference of 1.19 (31.24–30.05) points is even above statistical noise, given the large sample size of more than 1000 (which causes *p*-inflation), let alone the lack of MCID in their article. What is more, they did not even employ *p*-adjustment (correcting for false positives) for multiple comparisons, let alone dealing with the immense imbalance.

These and similar results call for action in psycholinguistics and in linguistics studies in general: reviewers and readers should be suspicious when they see results like these, even just by "eyeballing" the results, the magnitude of differences, and the lack of *p*-adjustment.

However, the novel recommendation in this book is to calculate and report MIDs for each and every difference in order to convince the readers to stay on the ethical side, and to avoid Type-I error (as in the study by Otero-López et al., 2021). The latter could have been avoided by reporting on effect sizes, specifically, partial eta-squared (η_p^2), as ANOVAs were performed in the study by Perea and Lupker (2003).

In my view, in addition to these, interindividual differences could have been considered in their studies (Perea & Lupker, 2003) by using, for example, linear

[1] Note: Non-CB denotes bon-compulsive buyers, whereas CB designates compulsive buyers.

mixed-effects models (e.g., Meteyard & Davies, 2020), which has been a common and default practice in psychology (Meteyard & Davies, 2020). Crucially, interindividual differences do play a role in MIDs.

The example of 7 s in Bourassa and Besner (1998) is a typical case of false positives (Type I error) in large samples, a topic we have already touched upon (e.g., Angst et al., 2017; Brezina & Meyerhoff, 2014; Kalinowski & Fidler, 2010; Lin et al., 2013; Sullivan & Feinn, 2012 *i.a.*).

Moreover, even marginally significant results are reported in Perea and Lupker (2003, p. 832): "The interaction between relatedness and type of prime was marginally significant [] $p<0.08$." I find this "marginal significance" utterly unacceptable given the small extent of differences in their experiments. I have reported this particular study to show the necessity of MIDs in the field of psycholinguistics also. These (not even) negligible "differences" undermine the practical significance of their research findings and showcase examples of Type I error in my opinion.

Finally, an arbitrary and small sample size of 18 participants (Perea & Lupker, 2003) and a difference of 7 ms (Bourassa & Besner, 1998) could have been avoided and spotted by the reviewers if the number of participants and a clear MID threshold had been prespecified in line with, for example, the IQWiG recommendations (IQWiG, 2023) or with good scientific practice.

One way of prespecification is preregistration of the study or simply following ethical scientific guidelines by writing an SAP before starting the study. In sum, these two studies just showcase the immediate need for MIDs or at least effect sizes in linguistics research.

Perea and Lupker (2003) write that "of course, one can never prove the null hypothesis, and the fact []," which is clearly not true given the existence of Bayesian factor (and other statistical methods) or the 95% Confidence Interval method outlined in IQWiG (IQWiG, 2023). Furthermore, the authors write that "In contrast, TL-final nonword primes and RL-nonword primes show only minimal evidence of being able to produce semantic/associative priming effects (around 3–5 ms in all cases)." In my mind, this statement is utterly unacceptable as a scientific claim. How can one contend that a random fluctuation of 3 ms is considered "minimal evidence," a term I have heard for the first time.

I would like to (re)emphasize in this chapter too that there are two scenarios in research designs: (i) the so-called pre–post measurements, sometimes referred to as within-subject design (i.e., the same group being measured two times), and (ii) a comparison of two independent groups, sometimes referred to as between-subject design (i.e., two different groups such as monolinguals and bilinguals). Note that computing MIDs for the two designs is different.

In the case of pre–post measurements (i), the calculation of MID is very easy: we take the baseline SD and half it (for instance, if we wish to calculate half an SD as MID in line with Norman et al., 2003, 2004). In other words, half an SD of the baseline SD will be our MID (if we opt for the most widespread 0.5 SD as MID), and if the increase or decrease exceeds half an SD (that is our MID), then we can claim that the extent of increase/decrease has practical/linguistic relevance/meaningfulness.

Consistent with Watt et al. (2021), the computation for (ii) is more difficult. We first have to calculate the pooled SD in line with Cohen (1988): $\sigma_{\text{pooled}} = \sqrt{([(\sigma 1_2) + (\sigma 2_2)] / 2)}$, where σ denotes the SDs of the two independent groups. After this, the MID is calculated: MID = 0.5 * σ_{pooled} (see Watt et al., 2021). In the latter formula, again, half an SD MID is shown.

Note that if another MID is to be used, then the SD or the pooled SD should be multiplied by, for instance, 0.3 * SD (e.g., in line with Eton et al., 2004), depending on factors such as expert opinions (who have an idea about the extent and magnitude of the effect) or preexisting studies in which MIDs have already been calculated for that context or phenomenon.

Linguistics traditionally does not utilize self-reported measures from individuals regarding, for example, their functional status or subjective experiences related to their health conditions, which pose a methodological problem. For instance, anchor-based methods cannot be used. The applicability of implementing MID-like metrics in these areas (e.g., practical or relevant difference), where direct patient input is absent has to be resolved to avoid false positives and unethical research practice such as "*p*-hacking" (e.g., Stefan & Schönbrodt, 2023).

The idea of establishing an MID in disciplines such as linguistics is challenging for another reason too: how do we choose the multiplier? Should it be 0.2 SD, 0.3 SD, 0.4 SD, or 0.5 SD? This problem also arises in other fields outside linguistics. To emphasize, my suggestion is to involve experts who can estimate the extent of the effect and/or consult preexisting literature in which MIDs have already been reported in a similar study.

Finally, linguistics could and should profit a lot from and accommodate MID defined as a measure of the practical relevance of a statistically significant difference. My recommendation is, consistent with Norman et al. (2003), that the widespread 0.5 SD MID should be used in linguistics as a benchmark for determining *linguistic meaningfulness*. The latter concept can be interpreted in the context of the impact of interventions in language-related disorders such as specific language impairment (SLI), among others.

For example, in therapeutic interventions, an improvement in *linguistic meaningfulness* (e.g., operationalized as *more coherent speech*) can indicate a significant clinical improvement, which can also be observed by the individual, the caregiver, the clinician, the parents, etc. If, however, no external "anchor person" (the mentioned individuals who can give feedback) is available, then only the distribution-based MID can be used.

Alternatively, if no "anchor person" is available (parent, teacher, caregiver, nurse, researcher, etc.), post-task comprehension tests in linguistic experiments can be used to measure how well participants understood the language: better comprehension scores might indicate improvements in *linguistic meaningfulness*; hence, the extent of the significant can be taken as exceeding the MID (i.e., linguistics meaningfulness), which can serve as proxy for the same.

Recall that De Vet et al. (2023), Engel et al. (2018), and Cook (2008) also articulate challenges in determining MIDs/MCIDs, such as *interindividual variability* and *subjectivity*. Given the latter two concerns, there should be a model that handles both concerns in linguistics too (beyond other disciplines).

One way of mitigating the problem is to divide the data into subgroups (strata) based on baseline characteristics such as age, gender, region, disease progression, disease severity (even in linguistics, see, e.g., specific language impairment), and other baseline scores (so-called *stratification*). Then, one would calculate the MID (e.g., half a SD of baseline scores) *for each subgroup separately*. In my mind, this straightforward method of stratification seems to be the easiest solution to approximate MIDs and a more personalized approach to MIDs. Though not in the field of linguistics but Bende et al. (2023) used stratification to compute the MIDs for males and females and adult and juveniles separately.

Baseline characteristics vary depending on the study. To name a few: gender, age, socioeconomic status (SES), race/ethnicity, educational level (highly important in the field of linguistics), employment status, income, medication use (some medicines can slow down response times in a psycholinguistics experiment, coffee intake (e.g., in an EEG experiment), hours slept (e.g., in an EEG experiment), cognitive function, and psychological and/or psychiatric comorbidities inter alia.

The Role of MID in Language Philosophy and Communication Sciences

While the MID concept has been examined mainly in the clinical field, its application in language philosophy and communication studies is still scarce. MID in the context of communication and language philosophy represents the smallest change in a variable that has value to a particular group of people or individuals. In *language philosophy*, MID can relate to *linguistic minimalism*, and the minimal level of variation needed for understanding, interaction, and persuasion to take place.

Within the *communication sciences*, MID is useful in assessing the effectiveness of speech, text, and media. It provides a basis to determine the level of efficacy in communication, that is, the level where *minimal changes to language result in drastically different levels of comprehension, engagement, or influence in social interaction*. This idea is pertinent in pragmatics, discourse analysis, semiotics, or where even the slightest changes in tone (e.g., tonal languages), phrasing, or phrasing mark a profound change in discourse.

A simple example from linguistics where a minimal change in language use can significantly affect comprehension or cause misunderstanding is the placement of a comma in written communication. For example, let us take a look at this the sentence "Let's eat, Grandma!." This sentence clearly means that the speaker is inviting their grandmother to eat. In contrast, the sentence "Let's eat Grandma!" (without the comma) suggests that the speaker is proposing to eat their grandmother! (sorry for the grotesque example).

This example demonstrates how a seemingly trivial punctuation mark (e.g., a comma) can drastically alter the meaning of a sentence, leading to potential misinterpretation. This principle extends to *spoken language* as well, where prosody (intonation and pauses) plays a similar role in comprehension.

Consider, for instance, the same spoken language phrases: "Let's eat, Grandma!" vs "Let's eat Grandma!." It is important, again, to note that the phrase does not contain a comma before it. That is one reason why the phrase is problematic in explaining, but simple when it comes to writing, since there is no conclusion without a change in tone or pause.

In all forms of speech, especially in public speaking, intonation and pitch changes in a speaker's voice (otherwise known as prosody in linguistics) greatly assist in clarifying the intended message.

With a slight pause after "eat" and a rising intonation on "Grandma," it becomes clear that the speaker is addressing Grandma. However, without a pause and with a flat or falling intonation, it might be perceived as a grotesque scenario (i.e., an alarming suggestion of cannibalism).

When it comes to tonal languages such as Thai, Punjabi, or Mandarin Chinese, it is immensely hard for language learners to decipher the meaning of a word because they cannot hear the tone properly. A tonal language is one in which the movement of a word or phrase's pitch affects its meaning.

The combination of MID with language philosophy and communication philosophy enables researchers to *measure the impact of small linguistic changes on human thought and behavior*. Such an understanding is beneficial, for example, in the case of rhetoric, translation, language therapy, and other branches of social work or in the understanding of minimal distinctions in language.

To start with, the *minimal language approach* by Diget (2022) proposes that the use of easily translatable and comprehensible texts can be employed to aid the understanding of people from different linguistic backgrounds. This approach fits within the MID concept on the premise that low language complexity changes greatly enhance understanding and communication effectiveness (Diget, 2022). Conclusively, MID is a relevant consideration for educators and communicators to determine the effectiveness of language use in relation to learners' comprehension.

However, unfortunately, no universally accepted criterion exists that differentiates minimal and nonminimal languages. Still, one can devise approximations based on different parameters: an example of criterion distinction is "Sufficiency of practical communication": "if a certain system enables communicating without any additional complexity, it is minimally sufficient" (Diget, 2022). In my opinion, a threshold for each of the studied components should have at least been proposed.

Another example comes from the spectrum of autism spectrum disorder (ASD). Communications practices of minimally verbal children, MID, are highly relevant. It has been shown that the language-focused therapies provided to children in this age range can be analyzed in terms of MID since it offers a means to assess whether or not the change in ability to communicate is meaningful and useful for the child's social interactions and development (Tager-Flusberg & Kasari, 2013; Saul & Norbury, 2020). Determining a specific MID provides practitioners the opportunity to measure the impact of different therapeutic strategies and *customize* them for profoundly verbal children (e.g., define an individualized MID).

Also, the MID concept can be utilized in the study of language disparities in qualitative research. Scholars engaging in research that includes participants from multiple language groups needs to account for MID to ensure that the changes observed in the use or comprehension of language are considerable enough. This is especially challenging in cross-language qualitative studies, where the impact of the meanings and their subtleties is often lost in translation, giving rise to data misinterpretation (Nes et al., 2010; Nasri et al., 2020). With the MID in mind, scholars may remain capable of ensuring the meanings and intent in their findings are not changed in ways that the participants' experiences are completely different from what is presented.

MID is also connected to the philosophy of language in the analysis of culture and communication. For example, the study of tonal languages draws attention to the role that pitch and intonation play in communication, which constitutes MID in linguistic expression. The slightest variations in tones can lead to drastic differences in meaning, which shows the need to consider such elements in language learning and communication activities (Dzokoto et al., 2016). When the MID in tonal languages is recognized, educators and linguists are able to better understand the intricacies of language acquisition and the role of culture in communication.

Furthermore, MIDs could be useful when attempting to formulate effective communication protocols within multidisciplinary collaborations, such as those between speech language pathologists and teachers.

In the area of philosophy of language, especially in relation to the Hungarian language, one of the pragmatic test examples is the "Háromutas Olvasásvizsgáló Teszt" (translated as "Three Way Reading Assessment Test") designed by Dankovics et al. (2017). This test is meant to measure the reading ability of people with aphasia, a language disorder resulting from brain damage (Dankovics et al., 2017). Its importance stems from the fact that it not only evaluates the technical aspects of reading but also the pragmatic and communicative aspects of language use.

The Three-Way Reading Assessment Test in Hungarian is guided by the theory developed by Ellis and Young regarding the integrated word recognition model. This model concentrates on the interactions between the different processes of reading like phonology, semantics, and pragmatics (Dankovics et al., 2017). The goal with this model is to assess the reading performance of a person to the fullest by focusing on his understanding of the text in relation to the sociocultural context, which is so critical for people suffering from aphasia. This is crucial because their ability to understand language in a practical way is often deeply affected by medical conditions (aphasia).

Apart from evaluating reading abilities, the test considers the sociocontextual aspects in the use of language. For example, the capability to derive meaning from certain circumstances constitutes contextually appropriate use of language. The Three Way Reading Assessment Test contains activities that require the participant to go beyond reading a text and interpret the underlying meanings and the social context of language (Dankovics et al., 2017). This component matches the concepts of language pragmatics.

Moreover, this development of the test illustrates more modern approaches in language assessment that involves assessing pragmatic skills together with other traditional language skills. The Three Way Reading Assessment Test by Dankovics et al. (2017) tries to fill this void through its novel approach to evaluating comprehension which is foundational in analyzing the performance of persons with language disabilities in pragmatic communication.

In addition, this case has repercussions that go beyond the clinical scope. The examination of the everyday problems people with aphasia encounter uses social constructs to further analyze the functioning of language. This observation corresponds to the social philosophical problem of language's functional essence and highlights that language is not a simple construction of symbols, but an active medium for communication that is shaped by the environment and societal practices.

To summarize, the "Háromutas Olvasásvizsgáló Teszt" is a standardized Hungarian language case of a pragmatic evaluation tool as it combines theoretical approaches from the philosophy of language and practical language analysis of reading. Its concentration toward the interaction of reading, comprehension, and pragmatic reasoning denotes the importance of the situation on the use of language, thus making it beneficial to researchers and practitioners in language evaluation. This test could benefit a lot from the concept of both anchor- and distribution-based MIDs.

Educational Sciences

Let us delve into the world of education sciences by taking a look at the study "Gender, Culture, and Sex-Typed Cognitive Abilities" by Reilly (2012). The study reports statistically significant gender differences, for example, in mathematical literacy (Cohen's $d = 0.13$) across OECD nations. However, from an MCID/MID perspective, the practical significance of this finding is minimal or close to zero in my mind, as this particular small effect size, for example, indicates that the observed significant differences do not translate into meaningful or noticeable differences in real-world educational outcomes or cognitive abilities. This calls into question the practical importance of the reported gender differences. The focus on small effect sizes, such as a Cohen's $d = 0.13$, could be seen as an overemphasis on differences that are trivial in practical terms or can even be due to Type I error given the large sample size of 480,405 (sic) students from across 65 nations.

Crucially, by highlighting these small differences, the study might mislead readers that they believe that these differences are more important than they actually are in practical terms. Also, which I find problematic, is that the study attempts to generalize findings across diverse OECD nations without sufficiently considering some factors such as different educational systems, teacher quality, school environment (Hanushek, 2002), SES, psychological factors, learner types (slow vs fast learners), the role of peers (Hanushek, 2002), or baseline scores in general that might mediate gender differences in cognitive abilities. A general MID would not take any of these into consideration.

In contrast, MCID/MID emphasizes context-specific thresholds for practical relevance. Furthermore, the failure to account for varying educational systems, cultural norms, and SES factors may limit the applicability of the findings and reduce their practical relevance. Moreover, the study does not establish clear thresholds for what constitutes a practically important difference (MID) in cognitive abilities.

I deem it is as a hiatus that could have been remedied by utilizing any MID. From an MCID/MID perspective, it is crucial to define what magnitude of difference would be considered meaningful in practical terms in the field. Without these thresholds/cutoffs, it is impossible or challenging at best to interpret the practical significance/importance of the reported differences. Experts in the field of educational sciences should form an expert panel and agree on the minimal extent of difference. Reporting small and statistically significant differences without a discussion of their practical relevance can lead to misinterpretation.

Educators, policymakers, stakeholders, and teachers might misinterpret these findings as indicative of substantial gender differences, potentially leading to misguided interventions or policies. This study shows the problem, even with effect sizes and the legitimacy of my writing the book on MID. Reilly (2012) presents statistically significant results along with effect sizes, with the effect sizes being close to zero. The use of some MID would have been crucial in my mind. Here are some specific citations from the article that point to either no effect or Type I error: "Overall, across non-OECD nations surveyed there was a statistically significant difference in science literacy favoring girls, **d=−.09** 95%CI [−.14, −.04], Zma=−3.44, p=.001," or "When both OECD and non-OECD nations were combined, there was a statistically significant difference in favor of girls, **d=−.04** [95%CI−.070,−.013] Zma=−2.84, p=.005.," or "Gender differences across PISA partner nations also favored males, Cohen's **d=.07** 95%CI [.02,.11], Zma=3.10, p=.001 although this difference was somewhat smaller" (bold emphasis mine, IF).

For example, for science literacy, the effect size is so small ($d = -0.09$) that it is either suggestive of a false positive result given the extremely large sample size or indicative of the possibility that while boys slightly outperform girls in these areas, the differences are minor and likely to be trivial in most educational settings. In my mind, these small differences do not justify major interventions, especially if other factors, such as societal or cultural play a larger role and remain uncontrolled.

In contrast, Hyde (2005) exercised a scientifically more rigorous approach and pointed out that gender differences in mathematics performance, for instance, were found to be negligible, with effect sizes (d) generally less than 0.10. This suggests that the differences between male and female students' performance in mathematics are *very small and hence not practically significant*. Still, the incorporation of an MID metric might have been crucial. By the way, Hyde (2005) is against gender differences based on negligible effect sizes. She contends that females and males are more similar than different on most psychological variables, for example, in contrast to the previous study by Reilly (2012), Hyde (2005) pointed out that 78%

of reported gender differences were small or negligible. Her research is very important because it calls for the application of MID in education research too. However, she does not address MID explicitly.

Sociology

Granovetter's seminal work from 1978 (1978) is centered around the dynamics of collective behavior and social movements using the framework of threshold models, similar to MCIDs/MIDs but with the model both incorporating one person and other people too: Granovetter (1978) introduced the concept of thresholds to model how *individual behaviors aggregate and culminate into collective actions*. Each individual has a threshold level of the number of others who need to adopt a behavior before they themselves will adopt it. This concept is essential in understanding how *small changes in the behavior of others can lead to large-scale social phenomena.*

The model is built on the idea that an individual will join a collective action when the proportion of others already engaged in the behavior exceeds their personal threshold. Mathematically, this can be expressed as a function where the probability of an individual joining a movement increases as more individuals join, with the individual joining only when the number of supporters reaches or exceeds their threshold.

The thresholds model explains why certain social movements or behaviors can spread rapidly once a critical mass is reached, while others do not. It provides insights into the dynamics of social influence and the spread of innovations or behaviors through social networks. A cutoff for a person is the proportion or number of other people who must engage in a behavior before that individual decides to join in.

This cutoff is personal and varies among individuals, in line with my conception of individualized MID, by the way. Granovetter (1978) often assumes a normal distribution of thresholds to simplify analysis, but other distributions can also be used depending on the specific context. Individuals make their decisions in sequence rather than simultaneously. The process begins with some initial adopters ("pioneers") whose participation starts the collective action. As more individuals join, the current level of participation will increase.

The point at which enough individuals have joined the collective action such that the process becomes self-sustaining is known as the critical mass. This is a major concept in the model as it determines whether a large-scale adoption will occur, changes, influencing others to join if their thresholds are met. The model reaches equilibrium when no additional individuals are willing to join the collective action. To illustrate, if individual 1 joins (because their threshold is 0.1, meaning they join if at least 10% of the group is participating), then the participation level becomes 0.1. This could trigger individual 2 to join because their threshold is 0.2.

In Granovetter's model (1978), *each individual has an individual threshold* that determines when they will join a collective action based on the behavior of others. Similarly, in clinical research, each patient has a *subjective and individualized threshold* (according to one view) that determines when they perceive a change in their health status as clinically noticeable (MCID). Granovetter's model (1978) shows how individual thresholds can lead to a *tipping point*, resulting in a large-scale social change.

In clinical trials, identifying MCIDs helps determine the point/threshold at which treatment effects are not just statistically significant but also meaningful to patients, potentially influencing clinical practice and health-care policies. The concept of critical mass in Granovetter's model (1978) can be analogized to a critical proportion of patients achieving their MCID. If a sufficient number of patients reach their MCID, the treatment may be deemed effective for the population (e.g., just think of COVID vaccination), similar to how a social movement gains momentum once a critical mass of participants is reached.

Moreover, in clinical practice, understanding MCIDs can influence sequential decision-making. For example, if early adopters of a treatment (*patients who perceive benefits at smaller changes*) report significant improvements, this could influence later adopters (*patients with higher MCIDs*) to try the treatment, akin to how early participants in a social movement influence others.

Although Mark Sanford Granovetter's work (1978) originates in sociology and social network theory, there are potential connections to the concept of MCID, particularly in the social sciences and educational contexts. For example, a small improvement in behavior or academic performance (even if below conventional statistical significance thresholds) might be considered minimally important if it helps bridge gaps between underperforming students and their peers.

Granovetter's (1978) theory also suggests that incremental changes may not seem immediately impactful, but they can have significant long-term effects. Similarly, MCID/MID focuses on detecting changes that, although small, might still be relevant to individuals. For instance, in educational settings, minor improvements in a student's performance, which may not result in a large effect size, can still be critically important for that student's engagement and long-term success.

References

Angst, F., Aeschlimann, A., & Angst, J. (2017). The minimal clinically important difference (MCID) raised the significance of outcome effects above the statistical level, with methodological implications for future studies. *Journal of Clinical Epidemiology, 82*, 128–136. https://doi.org/10.1016/j.jclinepi.2016.11.016

Bende, A., László, R., Faragó, S., & Fekete, I. (2023). An investigation of the possibilities for sex and age *determination of Eurasian woodcock (Scolopax rusticola L.) based* on biometric parameters using conditional inference trees and minimal important differences. *Journal of Vertebrate Biology, 73*(23068), 1–15. https://doi.org/10.25225/jvb.23068

Bourassa, D.C., Besner, D. (1998). When do nonwords activate semantics? Implications for models of visual word recognition. *Memory & Cognition 26*, 61–74. https://doi.org/10.3758/BF03211370

Brezina, V., & Meyerhoff, M. (2014). Significant or random?: A critical review of sociolinguistic generalisations based on large corpora. *International Journal of Corpus Linguistics, 19*(1), 1–28. https://doi.org/10.1075/ijcl.19.1.01bre

Cliff, N. (1993). Dominance statistics: Ordinal analyses to answer ordinal questions. *Psychological Bulletin 114*, 494–509. https://doi.org/10.1037/0033-2909.114.3.494

Cliff, N. (1996). *Ordinal methods for behavioral data analysis*. Mahwah, NJ: Lawrence Erlbaum.

Cohen, J. (1988). *Statistical power analysis for the behavioral sciences* (2nd ed.). NJ: Lawrence Erlbaum Associates. https://doi.org/10.4324/9780203771587

Cook, C. E. (2008). Clinimetrics corner: The Minimal Clinically Important Change Score (MCID): A necessary pretense. *Journal of Manual & Manipulative Therapy, 16*(4), E82–E83. https://doi.org/10.1179/jmt.2008.16.4.82E

Crosby, R. D., Kolotkin, R. L., & Williams, G. R. (2003). Defining clinically meaningful change in health-related quality of life. *Journal of Clinical Epidemiology, 56*, 395–407. https://doi.org/10.1016/S0895-4356(03)00044-1

Dankovics, B., Szilágyi, A., & Kóbor, A. (2017). Olvasási Készség Vizsgálata Afáziás *Betegeknél A Háromutas Olvasásvizsgáló Teszttel. Különleges Bánásmód - Interdiszciplináris Folyóirat, 2*, 87–98. https://doi.org/10.18458/kb.2017.2.87

De Vet, H. C. W., Terwee, C. B., & Mokkink, L. B. (2023). Minimal important difference. In F. Maggino (Ed.), *Encyclopedia of quality of life and well-being research*. Springer. https://doi.org/10.1007/978-3-031-17299-1_1813

Diget, A. (2022). Minimal language approach. *Scandinavian Studies in Language, 13*(1), 1–15. https://doi.org/10.7146/sss.v13i1.135070

Dzokoto, V., Senft, N., Kpobi, L., et al. (2016). Their hands have lost their bones: Exploring cultural scripts in two West African affect lexica. *Journal of Psycholinguistic Research, 45*, 1473–1497. https://doi.org/10.1007/s10936-016-9415-5

Engel, L., Beaton, D. E., & Touma, Z. (2018). Minimal clinically important difference: A review of outcome measure score interpretation. *Rheumatic Disease Clinics of North America, 44*(2), 177–188. https://doi.org/10.1016/j.rdc.2018.01.011

Eton, D. T., Cella, D., Yost, K. J., Yount, S. E., Peterman, A. H., Neuberg, D. S., Sledge, G. W., & Wood, W. C. (2004). A combination of distribution- and anchor based approaches determined minimally important differences (MIDs) for four endpoints in a breast cancer scale. *Journal of Clinical Epidemiology, 57*, 898–910. https://doi.org/10.1016/j.jclinepi.2004.01.012

Farivar, S. S., Liu, H., & Hays, R. D. (2004). Half standard deviation estimate of the minimally important difference in HRQOL scores? *Expert Review of Pharmacoeconomics & Outcomes Research, 4*(5), 515–523. https://doi.org/10.1586/14737167.4.5.515

Fekete, I., Schulz, P., & Ruigendijk, E. (2018). Exhaustivity in single bare wh-questions: A differential-analysis of exhaustivity. *Glossa: A Journal of General Linguistics, 3*(1), 96. https://doi.org/10.5334/gjgl.549

Granovetter, M. (1978). Threshold models of collective behavior. *American Journal of Sociology, 83*(6), 1420–1443. https://doi.org/10.1086/226707

Hanushek, E. A. (2002). Publicly provided education. In *Handbook of public economics* (Vol. 4). Elsevier.

Hyde, J. S. (2005). The gender similarities hypothesis. *American Psychologist, 60*(6), 581–592. https://doi.org/10.1037/0003-066X.60.6.581

IQWiG. (2023). *General methods version 7.0*. https://www.iqwig.de/methoden/general-methods_version-7-0.pdf

Jaeschke, R., Singer, J., & Guyatt, G. H. (1989). Measurement of health status: Ascertaining the minimal clinically important difference. *Controlled Clinical Trials, 10*(4), 407–415. https://doi.org/10.1016/0197-2456(89)90005-6

Kalinowski, P., & Fidler, F. (2010). Interpreting significance: The differences between statistical significance, effect size, and practical importance. *Newborn and Infant Nursing Reviews, 10*(1), 50–54. https://doi.org/10.1053/j.nainr.2009.12.007

Lin, M., Lucas, H. C., Jr., & Shmueli, G. (2013). Too big to fail: Large samples and the p-value problem. *Information Systems Research, 24*(4), 906–917. https://doi.org/10.1287/isre.2013.0480

Meteyard, L., & Davies, R. A. I. (2020). Best practice guidance for linear mixed-effects models in psychological science. *Journal of Memory and Language, 112*, Article 104092. https://doi.org/10.1016/j.jml.2020.104092

Meyer, D. E., & Schvaneveldt, R. W. (1971). Facilitation in recognizing pairs of words: Evidence of a dependence between retrieval operations. *Journal of Experimental Psychology, 90*(2), 227–234. https://doi.org/10.1037/h0031564

Mokkink, L. B., Terwee, C. B., Patrick, D. L., Alonso, J., Stratford, P. W., Knol, D. L., Bouter, L. M., & de Vet, H. C. (2010a). The COSMIN checklist for assessing the methodological quality of studies on measurement properties of health nature measurement instruments: An international Delphi study. *Quality of Life Research, 19*, 539–549. https://doi.org/10.1007/s11136-010-9606-8

Mokkink, L. B., Terwee, C. B., Patrick, D. L., Alonso, J., Stratford, P. W., Knol, D. L., Bouter, L. M., & de Vet, H. C. (2010b). The COSMIN study reached international consensus on taxonomy, terminology, and definitions of measurement properties for health-related patient-reported outcomes. *Journal of Clinical Epidemiology, 63*(7), 737–745. https://doi.org/10.1016/j.jclinepi.2010.02.006

Nasri, A., et al. (2020). Cross-language qualitative research studies dilemmas: A research review. *Qualitative Research Journal, 21*(1), 15–28. https://doi.org/10.1108/qrj-12-2019-0093

Nes, E. B., et al. (2010). Language differences in qualitative research: Is meaning lost in translation? *European Journal of Ageing, 7*(3), 313–316. https://doi.org/10.1007/s10433-010-0168-y

Norman, G. R., Sloan, J. A., & Wyrwich, K. W. (2003). Interpretation of changes in health-related quality of life: The remarkable universality of half a standard deviation. *Medical Care, 41*(5), 582–592. https://doi.org/10.1097/01.MLR.0000062554.74615.4C

Norman, G. R., Sloan, J. A., & Wyrwich, K. W. (2004). The truly remarkable universality of half a standard deviation: Confirmation through another look. *Expert Review of Pharmacoeconomics & Outcomes Research, 4*(5), 581–585. https://doi.org/10.1586/14737167.4.5.581

Otero-López, J. M., Santiago, M. J., & Castro, M. C. (2021). Big five personality traits, coping strategies and compulsive buying in Spanish University students. *International Journal of Environmental Research and Public Health, 18*(2), 821. https://doi.org/10.3390/ijerph18020821

Perea, M., & Lupker, S. J. (2003). Does jugde activate COURT? Transposed-letter similarity effects in masked associative priming. *Memory & Cognition, 31*(6), 829–841. https://doi.org/10.3758/BF03196438

Prasad, M. (2024). Importance of minimal clinically important difference in medical research and guideline development. *Journal of the Epidemiology Foundation of India, 2*(1), 01–04. https://doi.org/10.56450/JEFI.2024.v2i01.001

Reilly, D. (2012). Gender, culture, and sex-typed cognitive abilities. *PLoS One, 7*(7), e39904. https://doi.org/10.1371/journal.pone.0039904

Revicki, D. A., Cella, D., Hays, R. D., Sloan, J. A., Lenderking, W. R., & Aaronson, N. K. (2006). Responsiveness and minimal important differences for patient reported outcomes. *Health and Quality of Life Outcomes, 4*(70), 1–5. :. https://doi.org/10.1186/1477-7525-4-70

Revicki, D. A., Hays, R. D., Cella, D., & Sloan, J. (2008). Recommended methods for determining responsiveness and minimally important differences for patient-reported outcomes. *Journal of Clinical Epidemiology, 61*(2), 102–109. https://doi.org/10.1016/j.jclinepi.2007.03.012

Saul, A., & Norbury, C. (2020). Does phonetic repertoire in minimally verbal autistic preschoolers predict the severity of later expressive language impairment? *Autism, 24*(5), 1281–1290. https://doi.org/10.1177/1362361319898560

Stefan, A. M., & Schönbrodt, F. D. (2023). Big little lies: A compendium and simulation of *p*-hacking strategies. *Royal Society Open Science, 10*(2), 220346. https://doi.org/10.1098/rsos.220346

Sullivan, G. M., & Feinn, R. (2012). Using effect size-or why the P value is not enough. *Journal of Graduate Medical Education, 4*(3), 279–282. https://doi.org/10.4300/JGME-D-12-00156.1

Tager-Flusberg, H., & Kasari, C. (2013). Minimally verbal school-aged children with autism spectrum disorder: The neglected end of the spectrum. *Autism Research, 6*(6), 469–481. https://doi.org/10.1002/aur.1329

Watt, J. A., Veroniki, A. A., Tricco, A. C., et al. (2021). Using a distribution-based approach and systematic review methods to derive minimum clinically important differences. *BMC Medical Research Methodology, 21*, 41. (2021). https://doi.org/10.1186/s12874-021-01228-7

Chapter 13
Application of MID in Biology

Abstract This chapter focuses on the unique use of the concept minimal important difference (MID) in the biological field, stressing its importance in identifying significant differences that are biologically and ecologically trivial compared to those that are truly important. MID is largely ignored in biology, yet it serves as an important innovation in how researchers analyze findings from large datasets, particularly in situations involving small but statistically significant changes.

The application of MID in biology is relevant in two contexts: within-subjects (pre- and post-measurements) and between-groups (for instance, comparing males with females or adults with juveniles). Each scenario requires different biologically derived MIDs, most of which are computed using distribution-based approaches. MID can, therefore, be used as a benchmark beyond which observed differences are considered meaningful instead of just statistically significant.

This has been recently exemplified in ornithology, where the use of MID was adopted by researchers analyzing biometric data of thousands of birds in order to filter out irrelevant statistical noise or trivial differences. Despite having highly significant p-values, some differences did not exceed the MID and hence were biologically unimportant. This illustrates the usefulness of MID as a critical "sieve" in large-sample studies where reliance on statistical significance alone is completely inadequate.

Likewise, in forest ecology, scholars have applied several MID thresholds to study the effects of vegetation shifts on soil chemistry. To capture critical ecological changes, they computed MID at different values, for example, 0.2, 0.3, 0.4, and 0.5 standard deviations, in determining which ecological variables changed significatively—both in a statistical and ecological context. A good example is the pronounced increase in the lower tree layer, which was marked by considerable growth and surpassed all MID thresholds, corroborating its ecological relevance.

The scope of application of MID is vast, covering numerous biological areas, including neurobiology, where effect sizes inform differences attributable to sex in brain anatomy, and environmental biology, where MID can assist in establishing thresholds of ecosystem harm caused by pollutants or climate changes. MID is also applied within agriculture in distinguishing environment-driven vs biologically inherent differences observed within genetically modified crops.

I. Fekete, *Transforming Data into Informed Decisions across Clinical and Non-Clinical Domains*, https://doi.org/10.1007/978-3-032-02595-1_13

In conclusion, this chapter proposes integration of MID into biological study design frameworks, suggesting stratification to address age, sex, or regional dividing population variances. This tailored MID technique enhances the scope and interpretive depth of the results, affirming their basis in reality as opposed to mere hypothesis testing (p-values). Coupled with such innovation, MID can strengthen research in the biological sciences through improved methodological rigor, significance, and scientific impact.

Keywords Biology · Ecology · MID · Distribution-based approaches · Biological relevance · Pre–post-measurements · Two independent groups comparison · Case studies in biology

Interpreting Biological Relevance by Employing the Concept of MID

The application of MID in biology provides additional supporting evidence for biologically relevant *within- as well as between-individual differences*. Computing the distribution-based MID for these two concepts is different. Biologically relevant differences (e.g., between two groups) or changes in biological datasets (e.g., pre- and post-measurements) can have *biologically meaningful* implications for understanding biological processes (e.g., differentiating between adult and juvenile animals based on biometric parameters, see, e.g., Bende et al., 2023). "The focus of research should be on […] what data tell us about the magnitude of effects, the practical significance of effects, and the steady accumulation of knowledge" (Kirk, 2003, p. 100).

This chapter explores the innovations in MID determination within the field of biology. To reiterate, there are two scenarios: (i) *pre–post*-measurements (i.e., the same group being measured two times), which we called within-subject differences, and (ii) a comparison of *two independent groups* (i.e., two different groups such as male and female individuals), which we called between-subject differences at the beginning of this chapter. We are going to discuss the computation for both scenarios.

In biology, the use of statistics is heavily dominated by the use of null hypothesis significance testing (NHST), a concept that brings with it many unrecognized challenges we have already touched upon in the book. The most critical of these is that NHST lacks two key central pieces: (1) the magnitude of an effect in question and (2) how precise the estimation is concerning the magnitude of the effect calculated. As the majority of biologists should, in principle, be interested in biological reality, which is likely to be captured by the magnitude of an effect (not its statistical significance). Nakagawa and Cuthill (2007) suggest that all biological journals present effect sizes alongside their confidence intervals (CIs). The application of both the effect size statistic and its confidence interval allows for better evaluation of the relationship between data and p-values, irrespective of significance level.

Nakagawa and Cuthill (2007) offer solutions to four key technical limitations concerning the computation of effect size, confidence intervals (CIs), and the respective citation: (1) when there are covariates, (2) when there is potential bias in estimating effect size, (3) when observing a non-normal error structure of the data and/or variances, and (4) in the presence of nonindependent data. Although the consensus is that effect sizes are difficult to interpret, they offer some guidance to the researchers.

Also, the constant use of effect sizes would help researchers analyze their findings in relation to past studies and thus offer incorporate findings into subsequent meta-analyses that have, rather remarkably, become the default method of quantitative review in biology (Nakagawa & Cuthill, 2007).

Nakagawa and Cuthill (2007) explain all the details of two classes of effect size statistics, which are dimensionless, and thus d and r (for d statistics as standardised mean difference and r statistic as correlation coefficient) because they are prevalent across most study designs and also since their computations are needed in meta-analysis.

Nakagawa and Cuthill (2007) offer solutions to four key technical limitations concerning the computation of effect size, CIs, and the respective citation: (1) when there are covariates, (2) when there is potential bias in estimating effect size, (3) when observing a non-normal error structure of the data and/or variances, and (4) in the presence of non-independent data. Although the consensus is that effect sizes are difficult to interpret, we offer some guidance to the researchers. Nakagawa and Cuthill (2007) aim to transform the biological sciences with their unrefined statistical approaches by providing the "Proposals for Reform" alongside serving as a foundational instructional guide working within biological statistics.

Case Studies in Biology

Most recently, to the best of our knowledge, for the first time in biology, Bende et al. (2023) have employed MIDs to filter *statistically significant but trivial differences* in the field of ornithology on five biometric parameters including body weight, body length, wing length, tail length, tarsus length, and bill length of woodcocks in Hungary, Europe.

Importantly, Bende et al. (2023) used stratification (Tables 3, 4, 5, and 6 in their article) to refine the MID they employed. Sex (female/male) and age (adult/juvenile) were the two factors. There is another issue as to why they resorted to MIDs. Namely, they report findings from a very large database containing 13,226 individuals, which could give rise to Type I errors and practically irrelevant differences.

Just imagine! Lin et al. (2013) would call the abovementioned sample an "extraordinarily large sample" (Lin et al., 2013), as it exceeds 10,000 observations. They mention a researcher (unnamed), who solely relied on statistical significance in a sample of 10,000 observations without even reporting on the effect size.

Crucially, Lin et al. (2013) add that in extremely large samples, p-values approach zero very rapidly (as I have already mentioned in the book multiple times). Lin et al. (2013) point out that if the p-value threshold is used to determine significance, researchers will be misled into asserting that there is evidence supporting their conclusions; however, these findings lack practical importance.

Lin et al. (2013) discovered a not insignificant number of papers that relied solely on low p-values and regression coefficients as the primary basis for evidence confirmation for their hypotheses. Lin et al. (2013) claim that there is a p-value issue in extremely large samples, explaining them with the example of more than 300,000 camera sales on eBay. Lin et al. (2013) contend that solving the p-value paradox will enhance the trustworthiness of large sample research.

Returning back to an even bigger sample, in every comparison, Bende et al. (2023) were interested in the differences between two independent groups such as males and females or adult and juveniles. These five biometric parameters served as dependent variables, and they compared adult female and juvenile female woodcocks in big data (see Table 6 in Bende et al., 2023). Bende et al. (2023) used Tukey HSD post hoc tests to filter Type I error (false positive results). However, even p-adjustment does not help in filtering out biologically relevant differences in the terms of Nakagawa and Cuthill (2007).

When body sizes of adult female and juvenile male individuals were compared during the spring migration of woodcock in Hungary, significant differences were revealed even after adjusting with Tukey HSD-adjusted p-values. It is not surprising given that large samples yield significant differences (e.g., Brezina and Meyerhoff, 2014; Farivar et al., 2004; Kalinowski and Fidler, 2010; Lin et al., 2013; Sullivan and Feinn, 2012). However, some of the differences were below the most liberal MID that they employed. Note that measurements are given in millimeter (mm), except weight in gram (g).

To compute MIDs, Bende and colleagues (2023) kept the MID as low as possible to capture any small potentially biologically relevant difference. Their aim was to adopt the 0.2 SD criterion (e.g., Samsa et al., 1999; Mouelhi et al., 2020), in order to detect even *minimal biologically relevant differences*.

To compute the MIDs, they first calculated the pooled SD from the two independent groups (e.g., female and male or adult and juvenile) consistent with the formula in Watt et al. (2021). After this, Bende and colleagues (2023) compared the MIDs to the estimated differences (δ) to examine if the δ values are below or above the MIDs.

For the simplified computation of the pooled SD, they used the formula: $\sigma_{\text{pooled}} = \sqrt{}([(\sigma 1_2) + (\sigma 2_2)] / 2)$, where σ denotes the SDs of the two independent groups. This is the formula most commonly used with Cohen's *d* for independent samples. However, the correct σ_{pooled} is = $\sqrt{}[((n1 - 1)s1^2 + (n2 - 1)s2^2) / (n1 + n2 - 2)]$ for two independent groups. Regarding the notation, n1 and n2 denote the sample sizes of the two groups, s1 and s2 represent their respective standard deviations, and $s1^2$ and $s2^2$ indicate the variances of the two groups. This type of "liberal" MID (in the sense that it is low) was computed using the following formula: MID = 0.2 * σ_{pooled} (see Watt et al., 2021). The pooled SD that Bende et al. (2023) used is not wrong, it is just a simplified (or unweighted) version of pooled SD in the context of their study.

Bende et al. (2023) deliberately opted for this simple formula in Cohen (1988), as research has demonstrated that there is no clear benefit over using the more complex formula of computing pooled SD rather than the simpler one, i.e., Cohen's alternative formula (e.g., Gonick and Smith, 1993). The more complex formula for two independent groups (such as body weight of males and females) can be found in Borenstein et al. (2021, p. 22) (Table 13.1).

It can be concluded that that the estimated difference in bill length (see the last row in the table)—albeit above the MID—is most probably statistical random variation, and a stricter MID such as the 0.5 SD MID (e.g., Norman et al., 2003) would not indicate it as a biologically relevant difference with some biological implication.

Moreover, even the estimated difference of 0.98 mm is too small (see the Δ column in Table 13.1), being in the negligible range (although highly significant after Tukey HSD correction). The fact that the difference is highly significant even after Tukey HSD correction casts doubt on the reliability of post hoc corrections such as the Holm–Bonferroni correction in big data in general. This legitimates further the employment of MIDs whenever possible as *additional supporting evidence* to *enhance the quality of the results*.

Most recently in a similar study, MIDs have also been used in the field of forest ecology (Sokołowski et al., 2024). Sokołowski and colleagues examined the effects of the expansion of *Carpinus betulus* (CB; European hornbeam) in the Białowieża Forest in Central Europe, on soil properties under changing climatic conditions. Their findings demonstrate that the increase in *C. betulus* cover has led to significant changes in soil chemistry, including increases in nitrogen content, soil pH, and exchangeable cations. Overall, their research deals with how forest vegetation changes due to climate change and as to how that can have profound impacts on soil biogeochemical processes.

Sokołowski et al. (2024) employed even four different distribution-based MIDs using the formula by Watt et al. (2021) to examine the *ecological significance* of changes in *C. betulus* (CB) cover and its effects on soil properties in forests. They

Table 13.1 The table is taken from Bende et al. (2023), emphasizing both statistical significance (after Tukey HSD-adjustment) and MID for the investigated biological parameters. The table provides a comparative analysis of body measurements between adult females and young males of a species. Whenever the MID is exceeded, Δ, designating estimated difference, is in bold. Trivial differences are biologically irrelevant

	Adult female	Juvenile male			
Body measurements	Mean (± SD)	Mean (± SD)	Δ	*P* (Tukey HSD)	MID (0.2 SD)
Body weight (g)	320.0 ± 29.8	310.7 ± 28.2	**9.40**	$p < 0.001$	**5.80**
Body length (mm)	341.1 ± 15.0	336.7 ± 15.9	**4.33**	$p < 0.001$	**3.09**
Wing length (mm)	202.5 ± 19.0	199.5 ± 17.1	3.30	$p = 0.02$	3.61
Tail length (mm)	84.8 ± 8.5	83.7 ± 8.6	1.10	$p = 0.23$	1.71
Tarsus length (mm)	38.3 ± 3.6	38.0 ± 3.7	0.26	$p = 0.91$	0.73
Bill length (mm)	73.4 ± 4.1	72.5 ± 4.0	0.98	$p < 0.001$	0.81

used MIDs to identify the smallest changes in ecological variables, such as nitrogen content and soil pH that are *ecologically meaningful*/of *relevance*. Their study also pointed out that these changes, which exceed the MIDs in soil chemistry, are crucial for understanding the impact of CB expansion on forest ecosystems. Crucially, both studies (Bende et al., 2023; Sokołowski et al., 2024) called these thresholds MIDs, indicating that they adopted the same concept from the clinical field.

For the sake of easier understanding of the variables in the table, the following abbreviations have been adopted by Sokołowski et al. (2024): upper tree layer [A1], lower tree layer [A2], understorey layer [B], light [L], moisture [W], trophic [Tr], and soil acidification [R] conditions. For the MIDs, let us take a look at the table in Sokołowski et al. (2024) (Table 13.2).

SD criterion: 0.2; 0.3; 0.4 and 0.5, Δ = past–present (the difference), pooled SD = $\sqrt{([(SD_{past}^2) + (SD_{present}^2)]/2)}$. "*Carpinus betulus* covered in tree (A1, A2) and shrub (B) layers in 1962 (past) and 2013 (present) with MID analyses. Differences between present and past are based on the Wilcoxon test (P-value given)" (table cited from Sokołowski et al., 2024).

As we can see, Sokołowski et al. (2024) employed four distribution-based MIDs: MID of 0.2 SD, MID of 0.3 SD, MID of 0.4 SD, and MID of 0.5 SD to determine *ecological relevance/meaningfulness*. Second, they computed the simple pooled SD ($SD_{pooled} = \sqrt{([(SD_{past}^2) + (SD_{present}^2)]/2)}$), as they had two independent groups ("past" and "present"), where the two symbols represent the SDs of the two independent ones. It can be observed in the table that lower tree layer [A2] "stood the test" of all distribution-based MIDs, providing the strongest piece of evidence for the *ecological relevance/meaningfulness* of this variable.

Because this book discusses the use of MIDs in various disciplines, I do not mean to discuss different definitions of meaningfulness/relevance (*ecological, biological relevance,* etc.). Crucially, a difference of 17.36 (a difference of

Table 13.2 The utility of various MIDs in Sokołowski et al. (2024) who measured across two separate times. A1, A2, B, L, W, Tr, and R are seven main parameters they examined. The table also demonstrates the difference (Δ) between the past and present values. Sokołowski et al. (2024) went as far as calculating four levels of MIDS that provide reference points to assess the practical relevance

Tree layer	A1	A2	B	L	W	Tr	R
Past	11.10	6.84	2.85	2.94	3.05	2.87	3.13
SD	23.95	17.36	8.91	0.50	0.28	0.45	0.37
Present	4.90	24.20	6.64	2.82	3.30	3.06	3.22
SD	15.10	26.13	9.42	0.55	0.22	0.44	0.32
Δ	−6.20	17.36	3.79	−0.13	0.25	0.19	0.08
P-value	0.05	<0.001	0.01	0.24	<0.001	<0.001	0.10
Pooled SD	26.22	25.35	11.12	0.64	0.32	0.55	0.43
MID 0.2 SD	**5.24**	**5.07**	**2.22**	**0.13**	**0.06**	**0.11**	0.09
MID 0.3 SD	7.87	**7.61**	**3.34**	0.19	**0.10**	**0.16**	0.13
MID 0.4 SD	10.49	**10.14**	4.45	0.25	**0.13**	0.22	0.17
MID 0.5 SD	13.11	**12.68**	5.56	0.32	**0.16**	0.27	0.22

past–present) exceeds even the strictest 0.5 SD MID-criterion (e.g., Farivar et al., 2004) of 12.68 (for lower tree layer [A2]), which points to strong *ecological relevance/meaningfulness*. This provides additional supporting evidence.

I regard it as important to mention that Sokołowski et al. (2024) treated "past" and "present" as separate groups rather than the same group being measured two times. It is obvious because they are not the same individuals or observation units. Accordingly, Sokołowski et al. (2024) employed the right formula to compute MIDs. Another argument for treating the data as two independent groups is that data from the present are not contingent on data in the past. In other words, we are not dealing with a pre–post-measurement.

Without MIDs, it is challenging to determine whether the variations observed are of *practical, biological, or ecological importance* or if they are merely irrelevant statistical fluctuations. I contend that biological relevance should be part of the study design, with biological relevance designating a *notable and genuine impact on the environment or individuals*.

Future studies should establish MIDs in various fields of biology too or relevant thresholds that can guide the interpretation of data. This would involve determining what changes in population metrics are considered *ecologically or practically significant*.

Likewise, Abramov and Baryshnikov (2000) did not use any MIDs at all, *which resulted in the acceptance of biologically irrelevant differences as per expert opinion* (personal communication with Dr. Attila Bende, University of Sopron, Hungary). MIDs should have been used to highlight biologically relevant differences in skull sizes in millimeters (see Abramov and Baryshnikov, 2000, p. 371).

The study by Abramov and Baryshnikov (2000) is intended to be presented as an example for the need of MID in the field of biology and not just in the clinical field. Abramov and Baryshnikov (2000) reported skull sizes in millimeters from various countries, but the sample sizes show much variation: one (!) skull from China, two (!) skulls from Malta, two (!) skulls from Crete, and 30 skulls from England. The differences between skull sizes are so small that at times biological relevance seems questionable as per the mentioned expert opinion.

My simple solution proposed in this book for the computation of MID in the field of biology too when comparing biometric parameters such as body weight, body length, wing length, tail length, tarsus length, or bill length is the most widespread criterion of half an SD (0.5 SD of the baseline measurement) adopted from the clinical field (e.g., Norman et al., 2004).

I suggest that the formula by Cohen (1988) should be used in the case of *two independent groups* to compute pooled SD. (In the case of a pre–post-measurement—uncomparable to the formula for two independent groups—0.5 SD of the baseline measurement should be the MCID/MID). MCID/MID can be computed by multiplying the pooled SD by 0.5. I claim that anchor-based methods are not helpful in the field of biology because they are just not available. To put it simply, we cannot "ask the animals or skulls" if the estimated difference is relevant or not, as in the case of patients (PROs).

I further recommend reporting effect sizes and Cis, in line with Nakagawa and Cuthill (2007), to foster research findings. As a result, we can ensure that the research findings will have biological relevance beyond statistical significance.

In sum, the computation of MID (or MCID) for two different (independent) groups is as follows. First, the means and SDs for each group should be calculated. Then, the MID can be defined using the pooled SD. A common choice is 0.5 times the pooled SD in line with, for example, Farivar et al. (2004) or Norman et al. (2003).

Note that this threshold of 0.5 SD can vary depending on the context, indication (medical condition), and field of the research (along with a bunch of other factors such as patient population, individual differences, and patient baseline characteristics). For example, if employing 0.5 times the pooled SD as the MCID/MID, then it is computed as follows: $MCID/MID = 0.5 \times SD_{pooled}$.

Another example of the use of MID is the work of Hardy and colleagues (Hardy et al., 2017), who evaluated physical relevance in scientific evaluations. They highlight the need for stating what is a difference that is biologically relevant before any studies are conducted, which is important to ensure that the findings are not just statistically significant, but also significant from a biological perspective. They claim that, at a minimum, specification of agents, effects, subjects, and conditions is necessary for effectively framing assessment questions.

A biological system can "respond" to some exposure, or biomarkers can be studied if they exceed some threshold. Hardy et al. (2017) deal with these issues and come up with some general nonspecific guidelines. But crucially, they emphasize that biological relevance should be assessed by an assessor and that evidence should be weighted. In their words, "[] relevant data have been collected, the biological relevance of the effects in each data set should then be appraised." They conceptualize biological relevance by stating that it should be based on an expert judgment and that the magnitude of the effect should be evaluated (Hardy et al., 2017).

In a similar trend, Parks and Beiko (2010) also separate biological significance from statistical significance in metagenomic analysis. They suggest that confirming the existence of differences that are statistically significant is irrelevant from a biological standpoint. Their work stresses the importance of considering MID as they focus on effect sizes and confidence intervals in data analysis. Likewise, by concentrating on the true ecological or taxonomic basis of observed phenomena, researchers are likely to quantify the biological importance of the phenomenon better.

In agricultural studies, Venkatesh et al. (2014) provide an example of utilizing MID in the evaluation of genetically modified crops and the studies surrounding them. These specialists report that significant alterations within the composition of crops usually result from the surrounding environment rather than biological changes in the organism, which is not biological in nature. This is congruent with the MID concept in that it seeks to prove whether biogenetically modified differences are indeed encouraging enough to create questions or conduct further investigations. They call attention to the one-sided paradigm in which GM crops are consumed without conducting sufficient studies of biogenetically modified organisms, in which MID needs to be met.

Moreover, the study by Lewis et al. (2023) concerned with the structures of brain networks attempts to show profound biological disparities between sexes using effect sizes. Their results indicate that certain brain networks, more salient for one gender than for the other, exist. These findings are important steps toward understanding sex-based biological differences. This example demonstrates the use of MID in neurobiology in which it is necessary to use effect sizes to measure the significance of difference observed in brain activities.

The idea of biological relevance is just as important in clinical work, Worsfold et al. (2004) point out. They elaborate on *reference change value* (RCV), which is used to track important changes in biomarkers during the treatment of osteoporosis. The RCV is set to define the threshold pertaining to the smallest alteration that would be indicative of adequate biological response in accordance with the MID. Adopting this decision is of particular importance in clinical practice as a means to safeguard the integrity of treatment outcomes statistically and biologically.

Innovations in MID Determination in Biology

MID in biology extends beyond clinical outcomes to include changes in biological markers, cellular processes, and ecological impacts. The determination of MID in these areas necessitates the integration of biological variability, ecological relevance, and, of course, statistical significance. Crucially, biological relevance should be distinguished from statistical background noise and random variations.

Methodological innovations in MID determination in biology, driven by advancements in high-throughput sequencing technologies, proteomics, and metabolomics, have revolutionized the ability to detect and quantify minute changes in biological systems. These technologies enable the identification of MID at the molecular level, offering insights into gene expression, protein abundance, and metabolic shifts.

Bayesian statistical methods offer a flexible framework for MID determination, incorporating prior knowledge and uncertainty. This approach is beneficial in biological research where data may be sparse, allowing for a more robust estimation of MID (de Vet et al., 2007).

Fourth, environmental biology employs MID, and it is used to assess the impact of pollutants, climate change, and habitat alterations on ecosystems. For example, determining the MID for pollutant concentrations that affect species diversity, or the health of aquatic systems is crucial for environmental protection and policymaking (Guttikunda & Goel, 2013).

In genomics and proteomics, MID can assist in discovering noteworthy alterations in the expression of genes or the concentration of proteins associated with diseases, developmental processes, or treatment responses. This has implications for personalized medicine and biomarker discovery too.

In agriculture, MID determination is vital for evaluating the effects of genetic modifications, fertilizers, and pest control measures on crop yield and health. This helps in optimizing agricultural practices and ensuring food security.

Drawing on ecology, molecular biology, and computational sciences will catalyze understanding MID and its implications (de Vet et al., 2007). However, ethical concerns related to the use of high-throughput technologies and data privacy are essential for responsible research and application of MID (Lin et al., 2013). The determination of MID in biology represents a significant innovation, with MID hitherto underemployed, enabling to detect and interpret meaningful and biologically relevant changes in biological systems.

There is one more aspect of MCID/MID that I have not emphasized enough in the book. It is not just the question as to which threshold to use but also what the findings really mean in the particular field. For example, a study by Negrier et al. (2021) on the dietary regimen of the woodpigeon (*Columba palumbus*) could be critiqued in these terms: it fails to explicitly discuss the *ecological importance* of various dietary differences. When we speak about any type of practical implication ("clinical significance," "linguistic relevance," etc.), we have to ponder about the meaning and implication of the difference.

To illustrate, while differences in acorn and corn consumption are statistically significant, it is unclear whether these changes bear *ecological importance and/or substantial enough to impact the woodpigeon's overall health*, reproduction rates, or long-term population dynamics among others. These I call the *real-world implications*, practically, the "biological relevance." Expert opinions could come in handy here.

Also, a refined MID approach, which I advocate in this book, would stratify MIDs by region, population, or season (in other fields by gender, age, etc.). This way, we could determine whether changes in diet are differentially important across these strata. Earlier, I called this approach as stratification or personalized/individualized approach. In a biological context, of course, we cannot talk about personalized/individualized approaches unlike in medicine or psychology.

Moreover, there is an absence of expert opinions or a report of literature-based thresholds in Negrier et al. (2021). The latter would bring the study to a higher level because we could compare the results to preexisting findings.

References

Abramov, A. V. & Baryshnikov, G. F. (2000). Geographic variation and intraspecific taxonomy of weasel Mustela nivalis (Carnivora, Mustelidae). *Zoosystematica Rossica, 8*(2), 365–402.

Bende, A., László, R., Faragó, S., & Fekete, I. (2023). An investigation of the possibilities for sex and age determination of Eurasian woodcock (Scolopax rusticola L.) based on biometric parameters using conditional inference trees and minimal important differences. *Journal of Vertebrate Biology, 73*(23068), 1–15. https://doi.org/10.25225/jvb.23068

Brezina, V., & Meyerhoff, M. (2014). Significant or random?: A critical review of sociolinguistic generalisations based on large corpora. *International Journal of Corpus Linguistics, 19*(1), 1–28. https://doi.org/10.1075/ijcl.19.1.01bre

Cohen, J. (1988). *Statistical power analysis for the behavioral sciences* (2nd ed.). Hillsdale, NJ: Lawrence Earlbaum Associates.

de Vet, H. C., Ostelo, R. W., Terwee, C. B., van der Roer, N., Knol, D. L., Beckerman, H., Boers, M., &; Bouter, L. M. (2007). Minimally important change determined by a visual method integrating an anchor-based and a distribution-based approach. *Quality of Life Research: an International Journal of Quality of Life Aspects of Treatment, Care and Rehabilitation, 16*(1), 131–142. https://doi.org/10.1007/s11136-006-9109-9, https://doi.org/10.1007/s11136-006-9109-9

Farivar, S. S., Liu, H., & Hays, R. D. (2004). Half standard deviation estimate of the minimally important difference in HRQOL scores? *Expert Review of Pharmacoeconomics & Outcomes Research, 4*(5), 515–523. https://doi.org/10.1586/14737167.4.5.515

Gonick, L., & Smith, W. (1993). *The cartoon guide to statistics*. New York: HarperPerennial.

Guttikunda, S. K., & Goel, R. (2013). Health impacts of particulate pollution in a megacity—Delhi, India. *Environmental Development, 6*, 8–20. https://doi.org/10.1016/j.envdev.2012.12.002

Hardy, A., et al. (2017). Guidance on the assessment of the biological relevance of data in scientific assessments. *EFSA Journal, 15*. https://doi.org/10.2903/j.efsa.2017.4970

Kalinowski, P., & Fidler, F. (2010). Interpreting significance: The differences between statistical significance, effect size, and practical importance. *Newborn and Infant Nursing Reviews, 10*(1), 50–54. https://doi.org/10.1053/j.nainr.2009.12.007

Kirk R. E. (2003). The importance of effect magnitude. In Davis S. F. (Ed.), *Handbook of research methods in experimental psychology* (pp. 83–105). Malden, MA: Blackwell.

Lewis, N., et al. (2023). Fine temporal brain network structure modularizes and localizes differently in men and women: Insights from a novel explainability framework. *Cerebral Cortex*. https://doi.org/10.1093/cercor/bhac462

Lin, M., Lucas, H. C., & Shmueli, G. (2013). Too big to fail: Large samples and the p-value problem. *Information Systems Research, 24*(4), 906–917. https://doi.org/10.1287/isre.2013.0480

Mouelhi, Y., Jouve, E., Castelli, C., & Gentile, S. (2020). How is the minimal clinically important difference established in health-related quality of life instruments? Review of anchors and methods. *Health and Quality of Life Outcomes, 18*(1), 136. https://doi.org/10.1186/s12955-020-01344-w

Nakagawa, S., & Cuthill, I. C. (2007). Effect size, confidence interval and statistical significance: A practical guide for biologists. *Biological Reviews, 82*(4), 591–605. https://doi.org/10.1111/j.1469-185X.2007.00027.x

Negrier, C., Fantinati, M., Jouglar, J. Y., Lyazrhi, F., Cohou, V., & Priymenko, N. (2021). Dietary regimen of the woodpigeon (Columba palumbus). *Journal of Animal Physiology and Animal Nutrition, 105*(2), 376–384. https://doi.org/10.1111/jpn.13409

Norman, G. R., Sloan, J. A., &; Wyrwich, K. W. (2003). Interpretation of changes in health-related quality of life: The remarkable universality of half a standard deviation. *Medical Care, 41*(5), 582–592. https://doi.org/10.1097/01.MLR.0000062554.74615.4C, https://doi.org/10.1097/01.MLR.0000062554.74615.4C

Norman, G. R., Sloan, J. A., & Wyrwich, K. W. (2004). The truly remarkable universality of half a standard deviation: Confirmation through another look. *Expert Review of Pharmacoeconomics & Outcomes Research, 4*(5), 581–585. https://doi.org/10.1586/14737167.4.5.581

Parks, D. H., & Beiko, R. G. (2010). Identifying biologically relevant differences between metagenomic communities. *Bioinformatics*. https://doi.org/10.1093/bioinformatics/btq041

Samsa, G., Edelman, D., Rothman, M. L., Williams, G. R., Lipscomb, J., & Matchar, D. (1999). Determining clinically important differences in health status measures: a general approach with illustration to the Health Utilities Index Mark II. *PharmacoEconomics, 15*(2), 141–155. https://doi.org/10.2165/00019053-199915020-00003

Sokołowski, K., Czerepko, J. & Gawryś, R. (2024). The effects of Carpinus betulus expansion on soil properties under changing climatic conditions. *Science Reports, 14*, 24662. https://doi.org/10.1038/s41598-024-73960-1

Sullivan, G. M., & Feinn, R. (2012). Using effect size-or why the P value is not enough. *Journal of Graduate Medical Education, 4*(3), 279–282. https://doi.org/10.4300/JGME-D-12-00156.1

Venkatesh, T. V., et al. (2014). Compositional analysis of grain and forage from MON 87427, an inducible male sterile and tissue selective glyphosate-tolerant maize product for hybrid seed production. *Journal of Agricultural and Food Chemistry, 62*. https://doi.org/10.1021/jf4041589

Watt, J. A., Veroniki, A. A., Tricco, A. C. et al. (2021). Using a distribution-based approach and systematic review methods to derive minimum clinically important differences. *BMC Medical Research Methodology 21*, 41. https://doi.org/10.1186/s12874-021-01228-7

Worsfold, M., et al. (2004). Assessment of urinary bone markers for monitoring treatment of osteoporosis. *Clinical Chemistry, 50*. https://doi.org/10.1373/clinchem.2004.037424

Chapter 14
MID in the Business Sector

Abstract The critical importance of minimal important difference (MID) within the realms of product management and business strategy lies in its ability to differentiate between changes that are meaningful versus negligible. MID delineates a line of differentiation where any action taken beyond that line crosses into the domain of consumer perceptions, organizational policies, or even hierarchical procedural frameworks.

Consider a scenario with a tech company that adds a new feature to their products that enhance battery life. If the enhancement does not meet the MID threshold, there is a high likelihood that consumers will not see it as a value addition. Businesses have the option of using visual analogue scales or Likert scales to measure customer satisfaction in order to evaluate such modifications. Obtaining MIDs by half a standard deviation, for example, increases the probability that the feedback obtained from customers is both statistically sound, practically useful, and perceivable to the customers.

The application of MIDs spans across almost all domains of a business. In consumer behavior, MIDs assist in determining the impact or effectiveness of marketing strategies by connecting the results from a campaign to relevant shifts in attitudes among consumers. In manufacturing, MID serves to guide technology investment by establishing the minimal improvements in operational efficiency required to justify a given expenditure. In finance, the application of MID can be seen in analyzing key performance indicators like return on investment and market share in determining whether a particular strategy necessitates further investment.

MID is also important in human resource management and employee relations. A small engagement metric improvement may not result in an increase in morale or productivity in the workplace until it surpasses the MID. This understanding prevents resource wastage due to insignificant changes.

The Google "blue experiment" of 2009 illustrates the issues that can arise from improper application of the MID concept. Without any bounds or attempts to calculate the MID, such conclusions run the risk of being misattributed or overestimated due to extremely large sample size alone with no control for confounding variables or an overreaching scope of measurement.

I. Fekete, *Transforming Data into Informed Decisions across Clinical and Non-Clinical Domains*, https://doi.org/10.1007/978-3-032-02595-1_14

In addition, MIDs serve critical purposes in software design, holistic supply chain management, and even in the ethics of business. For instance, guiding the critical MIDs identification of business ethics serves to deepen the bond of a company with its stakeholders, customers, and improves corporate governance.

Within the confines of health care and pharmacology, MID assists in patient care, patient management, treatment scheduling, and device allocation. It also enables policymakers and insurers to make informed decisions regarding coverage and tailored clinical intervention. MID is also essential within the company's marketing and competitive positions, as it is needed for regulatory submissions proving drug efficacy.

More importantly, strategic blunders, misallocation or even waste of resources, and stunted advancement possibilities due to MIDs not being integrated can affect decision-making in a negative way. Evidence supporting that performance and key organizational decisions are made based solely on data prompts is fundamentally justified when employing MID.

Keywords MID · Business · Patient management · Consumer behavior · Market research · Customer satisfaction · Customer surveys · Service quality · Business implications of MID

The Concept of MID in Business, Consumer Research, Consumer Behavior, Market Research, Customer Satisfaction, and Service Quality

In *product management*, minimal important differences (MIDs) could help determine the threshold for *noticeable improvements* in product features, for instance. Imagine that a tech company might use MIDs to decide whether an *increase in battery life or a certain reduction in charging time substantially* impacts *user satisfaction*. In statistical terms, a reduction in battery life would be the independent variable, and user satisfaction the dependent variable. It might be the case that there is a significant increase in battery time/life, but users do not perceive it as essential. We all know such examples from advertisements such as toothpastes or some electronic gadgets.

When using MIDs, my recommendation is employing a *visual analogue scale* (VAS) or a Likert scale to perform such a measurement. In my experience as a researcher, participants prefer VAS over Likert scales. VAS is basically a horizontal line; participants indicate on the line how satisfied they are with the product, for example. One advantage of these two feedback methods is that they can be administered online. First, there should be a sample-size calculation, and then one should run some traditional statistics. After this, the MID must be calculated based on, for example, half an SD (the strict one), to enhance the reliability of results. MID can be explained in simple terms to the customers, and, as a result, they will put their faith in the product. In contrast, if a dentist just announces verbally that he or she also uses that toothbrush that will not achieve the same target, namely, an increase in customers or customer satisfaction.

MID in the context of the business world also refer to the smallest change in a measurement that is perceived as important by consumers or stakeholders, influencing their behavior and decision-making processes substantially. In the business world, both the stakeholders and the customers are involved in decision-making. Imagine that you construct a survey and want to convince your colleagues that revenue or customer satisfaction has increased. MID is the best choice, in my judgment.

In consumer behavior studies, MIDs can be utilized to assess the impact of marketing strategies on customer satisfaction. For instance, when evaluating the effectiveness of a *new advertising campaign*, businesses can measure changes in consumer attitudes and behaviors against established MIDs. This approach not only enhances the understanding of consumer responses but also aids in refining marketing strategies to better suit consumer expectations and consumer trust if MID is presented to the consumers.

For example, in *manufacturing*, establishing MIDs for *production efficiency* can help identify the *minimum improvements* necessary to *justify investments* in new technologies or processes. Hence, businesses can allocate resources more effectively and optimize operational efficiencies.

MIDs can serve as thresholds in *financial performance* too for evaluating the success of strategic initiatives. For instance, when launching a new product, companies can set MIDs for *sales growth* or *market share increases* to assess the effectiveness of their marketing efforts. By comparing actual performance against these MIDs, enhanced decisions about future investments and adjustments can be made.

MIDs can enhance marketing effectiveness by providing a framework for evaluating the impact of various marketing tactics. As an example, marketers can evaluate campaign efficacy in order to refine results in meeting strategic objectives by setting MIDs for *customer acquisition costs* and *return on investment* (ROI) metrics (e.g., Podurgiel, 2024). This process can be accomplished by defining MIDs for so-called *key performance indicators* (KPIs) such as *customer acquisition costs* or *return on investment* (ROI) (e.g., Podurgiel, 2024).

Moreover, in *service quality assessments*, MIDs can provide valuable insights into the aspects of *service delivery* that matter most to customers. By identifying MIDs for various service attributes—such as *responsiveness*, *reaction time*, *reliability*, and *empathy*—organizations can prioritize improvements that will yield the greatest benefits in customer satisfaction. All these constructs can be measured using standardized tests, for example. Again, to stress, this focus on meaningful changes can lead to enhanced *customer loyalty* and *customer retention*. Moreover, they can be integrated in advertisements as objective evidence.

Let us talk about the negative aspects too, when MIDs are not used or misused. The absence of MID could result in either *overinvesting in marginal improvements or underestimating the impact of certain feature enhancements*. In human resources, understanding MIDs is essential for interpreting changes in *employee engagement*, *employee satisfaction*, or *performance metrics*. A small improvement in an engagement survey score might be statistically significant but not important/meaningful/relevant/substantial in terms of actual changes in *employee morale*, *productivity*, *performance*, or *for the customers*. Again, these constructs can easily be measured employing standardized tests. I should mention that a single-item scale can also be employed to measure these constructs.

Businesses need to understand what level of change in these metrics *genuinely* reflects an improvement in the workplace environment or performance. Determining the MIDs for these metrics helps in understanding the actual impact and direction of marketing strategies.

For instance, a slight but statistically significant increase in click-through rate might not be important/meaningful/relevant/substantial enough to justify the cost of a particular *advertising campaign*. Finally, in *financial analytics*, MIDs can inform *investment decisions*, such as understanding the significance of changes in market share or revenue growth.

An MID difference in revenue growth, for example, can help investors determine whether a company's growth is robust enough to warrant investment. Hence, the absence of an MID in these contexts leads to challenges in optimal decision-making. Businesses might either *overreact to insignificant changes* or *overlook significant trends*, as already mentioned.

Finding specific business examples where companies have used MID as a metric is challenging, as the concept is more commonly discussed in health-care and clinical research contexts, as you might have already perceived it.

Businesses, in their pursuit of growth and efficiency, often employ multiple metrics to measure performance, for example, *customer satisfaction* or *market trends*. However, without the integration of MIDs, these metrics may lead to misinterpreted data and misguided strategies. CEOs and team leaders might or might not have some knowledge of statistics, but when they do not employ MIDs as an extra measure, they might miss the optimal target.

MID can differentiate between trivial changes and those that are genuinely impactful to the business or the consumer. Its absence thus leads to suboptimal decisions—similar to a doctor treating a patient without knowing the threshold at which symptoms become clinically relevant to the patient.

This oversight can result in an allocation of resources toward initiatives that fail to meet the threshold of a meaningful impact, while neglecting areas where slight changes could yield significant benefits. For instance, a company might invest heavily in a marketing campaign that marginally increases customer satisfaction, overlooking minor product adjustments that could lead to substantial improvements in user experience and loyalty. It becomes paramount for businesses to incorporate MIDs into their strategic planning and assessment, in my view. Hence, they can maximize profit and avoid loss by making suboptimal decisions.

The lack of awareness regarding the metric of MIDs in the business industry can lead to a cascade of negative consequences, extending beyond suboptimal decisions and misguided strategies. These ramifications include, but are not limited to misinterpreted market trends, misallocation of resources (companies might allocate financial, human, and technological resources to areas that do not yield meaningful returns or improvements), overinvesting in marginal improvements, suboptimal investment decisions, misguided decisions, misinterpretations of trends, establishing false causality, and overreaction to insignificant changes.

One example is from 2009, when Google conducted a famous experiment where they tested 41 different hues of blue for their advertising links for a tool bar.[1] Their aim was to identify the exact shade of blue that would maximize user engagement as the dependent variable.

Results showed a statistically significant difference, with one particular shade of blue standing out significantly more than the other "blues" ("the greener blue"). Based on these findings, Google decided to implement this optimal shade of blue across all their platforms.

In the year of the study in 2009, Google had revenue from search advertising in the value of 22.7 billion US$ (IAB Internet Advertising Revenue Report). However, one concern from the perspective of MID is that the quoted 200.000.000 US$ profit, which is even less than 1%, casting doubt on the causality and impact of their "blue-experiment" and the MID.

I must say again that this 1% might reflect statistical fluctuation, rather than a genuine difference. If Google engineers had claimed that their results showed that a different shade of blue yielded less than 1% increase in revenue, one might think twice about venturing on this experiment in the first place, let alone believe in the causality and impact of the change of the hue. I am confident that the *extremely slight increase of 1%* is due to *p*-inflation (e.g., Brezina & Meyerhoff, 2014; Lin et al., 2013).

Brezina and Meyerhoff (2014), albeit not from the field of sociolinguistics, do not specifically focus on big data, but on the topic of false positives in big data. False positives are a concern in the Google-study too, for many reasons.

Unfortunately, I could identify other methodological problems with this experiment, for example, that users had different monitors, and the color sight of the participants was not measured and that other objective conditions of the experiment were not controlled (sample size, age, gender, reaction time, device type, browser type, etc.), let alone the time in milliseconds of presentation of the 41 blue stimuli.

Even if presentation time had been controlled for, every computer and processor are different. For example, a virus scanner can slow down reaction times. If you give more time to the participants, they might change their mind. Age, as already mentioned, was not controlled for either, although we know that the eye sight of senior citizens is different from the younger generation. Even ethnicity could play a role in color perception. In Russia, there are two words for blue ("синий" pronounced as "siniy" for dark blue and "голубой" pronounced as "goluboy" for light blue), whereas along the equator they "see" "grue" (the combination of green and blue).

But what is important for us from the perspective of MID is that they could not show that the human eye could distinguish between two different hues of blue. If the eyes could, they should have computed the MID in a post hoc manner. I am

[1] https://www.nytimes.com/2009/03/01/business/01marissa.html

confident that without an MID, given the large sample size, the finding is completely unreliable.

In conclusion, the failure to incorporate MIDs into business decision-making processes can have wide-ranging and profound effects on a strategy of company, operations, and ultimately, its success in the market.

Specific Examples for Applications

As an example, with respect to MID in software outsourcing, it is relevant to look at different business knowledge requirements for the project and how they affect MID. Mishra and Mahanty (2015), for example, indicate that one understands the minimal business knowledge differences that need to be dealt with during software development, particularly during developmental, reengineering, and maintenance phases. This entails that they have the knowledge of what is really required for more effective tailoring of the strategies crafted.

Also, in supply chain management, MID is important when measuring operational change performance. In line with Çalık's work (2020), it is essential to note that performance differences in the metrics must be benchmarked when considering supplier selection or resource allocation in decision-making in an integrated supply chain network (Çalık, 2020). If these minimal differences are reduced, the supply chain cost and service delivery will be maximally enhanced.

Similarly, the use of MIDs also extends to areas outside medicine, for example, business ethics. For instance, Heath et al. (2010) suggest that if important differences of ethics were recognized in business practices, it would enable better corporate governance and stronger relationships with the stakeholders (Heath et al., 2010). Being aware of what can be regarded as a *meaningful ethical difference* puts an organization in a position to meet the expectations of its stakeholders and thus, improve its image in the market (ultimately, giving them a competitive leverage).

Minimal Important Difference and Patient Management: The Business Implications

MID, also an essential indicator in patient management, has considerable business ramifications in the health-care sector. By concentrating on what people deem advantageous, health-care providers can deliver better patient-centered treatment. This strategy enhances patient pleasure and may also augment patient loyalty and trust in the health-care provider, which is essential in a competitive health-care landscape.

MCID/MID aids clinicians in rendering more informed treatment judgments. Comprehending MCID/MID enables health-care providers to more effectively evaluate the efficacy of various treatments. This may result in enhanced efficiency in *health-care resource utilization, strategic planning for sustainable performance* of health-care organizations, a *decrease in needless treatments*, and the *optimization of PROs*, all of which carry substantial business implications regarding *cost reductions* and *resource allocation*. Cost-effectiveness analysis (CEA) is essential in evaluating the costs and outcomes of various treatments, thereby assisting health-care professionals in making educated and efficient strategic decisions.

MCID/MID significantly influences health-care policy formulation and insurance coverage determinations. Insurers and policymakers frequently utilize MCID/MID to assess the efficacy of treatments for coverage justification. The MCID/MID offers a measurable assessment of treatment efficacy from the patient's viewpoint, potentially affecting reimbursement policies and coverage determinations, therefore influencing the financial aspects of the health-care sector.

In the pharmaceutical sector, MCID/MID is crucial for clinical trial design and medication approval processes. To illustrate, dimethyl fumarate (Tecfidera™), a compound for relapsing forms of multiple sclerosis (MS), from Biogen GmbH in Germany, is an example. MCID/MID affects regulatory approval and commercial access for novel pharmaceuticals. Pharmaceutical businesses can utilize it to showcase the efficacy of their goods, which are essential for obtaining market approval and achieving a competitive edge.

An example of a company that develops dossiers for German HTA is SmartStep Consulting GmbH (date: 2025), located in Hamburg, Germany. They have embraced the IQWiG's ≥ 15% threshold (IQWiG, 2023, p. 173); simultaneously, two very important criticisms have been expressed by SmartStep Consulting GmbH: (1) "this new threshold is no longer patient-centric, substituting a patient-centered approach with a technical one," and (2) "changes quantified in natural units or utilizing (at least theoretically) unlimited scales (e.g., 6-min walk test) are not addressed by the new regulation and still necessitate a validated threshold grounded in clinical data" (SmartStep Consulting GmbH[2]).

Health insurance companies and pharmaceutical businesses that adeptly employ MID in their communications might distinguish themselves in the marketplace. Crucially, health insurance companies want to decrease the costs, whereas pharmaceutical businesses want to do the opposite, i.e., increase the revenue. These two sectors can secure a competitive advantage; alas, they are "enemies" when it comes to costs.

In addition, MID functions as a crucial performance indicator for quality enhancement efforts in health-care companies. It establishes a target for health-care providers to enhance PROs, applicable for internal performance assessments, accreditation procedures, and external marketing efforts.

[2] https://www.smartstep-consulting.de/en/home-2/

In health care, the MID represents the smallest change in a PROs that patients perceive as beneficial (e.g., Bohannon, 2019). Still, there are refinements needed. For example, Bohannon (2019) demonstrates using grip strength as the PRO that more rigorous studies are required to determine the MCID for grip strength. With the aid of MID, health-care professionals can provide more accurate care tailored to patients' actual needs and preferences.

The realization of the significance of MIDs has several consequences for health-care organizations. For example, in the case of assessing a new treatment protocol, healthcare providers may find it useful to look at PROs and see if the changes in them are significant in the eyes of patients, especially in relation to widely accepted MIDs. While this strategy improves the quality of care, it also improves the loyalty and trust level of patients, which are vital for the success of health-care organizations.

In addition, the combination of MIDs and patient management techniques can significantly enhance the efficiency and, importantly, cut down costs. The interventions that result in significant PROs enable the health-care organization to economize and allocate resources more effectively, thereby minimizing wasteful spending (Johnston et al., 2015).

MacKay et al. (2019) question the confidence of WOMAC (the test they used) and call for a refinement of the MCID. Not just the patient, but also achieving a desired level of business effectiveness through clinical effectiveness is important toward maintaining competitiveness in the world of health care.

Overall, the relevance of MID in the context of patient management has vital business dimensions—patient satisfaction, treatment efficiencies, optimal allocation of resources, policies on health care, drug design and development, market competition, and quality improvement activities.

References

Bohannon, R. W. (2019). Minimal clinically important difference for grip strength: A systematic review. *Journal of Physical Therapy Science, 31*(1), 75–78. https://doi.org/10.1589/jpts.31.75

Brezina, V., & Meyerhoff, M. (2014). Significant or random?: A critical review of sociolinguistic generalisations based on large corpora. *International Journal of Corpus Linguistics, 19*(1), 1–28. https://doi.org/10.1075/ijcl.19.1.01bre

Çalık, A. (2020). An integrated open-loop supply chain network configuration model with sustainable supplier selection: Fuzzy multi-objective approach. *SN Applied Sciences, 2*(5), 1–15. https://doi.org/10.1007/s42452-020-2200-y

Heath, J., Moriarty, J., & Norman, W. (2010). Business ethics and (or as) political philosophy. *Business Ethics Quarterly., 20*(3), 427–452. https://doi.org/10.5840/beq201020329

IQWiG. (2023). *General methods version 7.0.* https://www.iqwig.de/methoden/general-methods_version-7-0.pdf

Johnston, B. C., Ebrahim, S., Carrasco-Labra, A., Furukawa, T. A., Patrick, D. L., Crawford, M. W., Hemmelgarn, B. R., Schunemann, H. J., Guyatt, G. H., & Nesrallah, G. (2015). Minimally important difference estimates and methods: A protocol. *BMJ Open, 5*(10), e007953. https://doi.org/10.1136/bmjopen-2015-007953

Lin, M., Lucas, H. C., Jr., & Shmueli, G. (2013). Too big to fail: Large samples and the p-value problem. *Information Systems Research, 24*(4), 906–917. https://doi.org/10.1287/isre.2013.0480

MacKay, C., Clements, N., Wong, R., & Davis, A. M. (2019). A systematic review of estimates of the minimal clinically important difference and patient acceptable symptom state of the Western Ontario and McMaster Universities Osteoarthritis Index in patients who underwent total hip and total knee replacement. *Osteoarthritis and Cartilage, 27*(10), 1408–1419. https://doi.org/10.1016/j.joca.2019.05.002

Mishra, A., & Mahanty, S. (2015). Business knowledge requirements and onsite offshore work division in Indian software outsourcing projects. *Strategic Outsourcing: An International Journal, 8*(3), 212–230. https://doi.org/10.1108/SO-10-2014-0025

Podurgiel, R., Piscitellit, D., & Denegar, C. (2024). Challenges in applying minimal clinically important difference: A critical review. *International Journal of Rehabilitation Research, 47*(1), 1–10. https://doi.org/10.1097/MRR.0000000000000613

Chapter 15
Economics, Economic Policy, Environmental Economics, Finance, and Health Economics

Abstract Bringing the minimal important difference (MID) into economic reasoning allows decision-makers to step beyond mere numbers and ponder whether a given policy affects people substantially. A drop in the unemployment rate that shows up as statistically solid might not be noteworthy at all. Benchmarks are often absent from cost–benefit analyses, which impacts reports, ultimately leading to wrong decision-making.

The chapter discusses the difference between statistical significance and economic significance. Also, differences-in-differences (DiD) analysis is discussed. DiD has become a staple in policy circles because it pairs simple pre–post comparisons while controlling for confounders.

In public-health and environmental–economics circles, the notion of willingness to pay (WTP) can be related to the MID. Both numbers aim to answer a simple question: what small improvement "feels" worth the money. That makes WTP estimates a staple in cost–benefit analyses.

Keywords MID · Economics · Health economics · Environmental economics · Statistical significance · Economic significance · Differences-in-differences analysis · DiD · Economic policy

Economics and MID

Incorporating MID into economics gives policymakers and stakeholders a chance to see if a policy change has any meaningful impact. For instance, if the unemployment rate decreases, it may be statistically significant in a large sample but may not in any way justify a policy change (the decrease may continue unchanged its policy of not adjusting the duration or amount of unemployment benefits). If this difference exceeds the established or (post hoc calculated) MID, which has not been established in most studies or programs, it will indicate whether the policy change could have a meaningful impact on the population under investigation.

I. Fekete, *Transforming Data into Informed Decisions across Clinical and Non-Clinical Domains*, https://doi.org/10.1007/978-3-032-02595-1_15

MID can equally be used to assess the effectiveness of certain policies or expenditure cuts. For example, when looking at the impact of a new tax system on income of households, economists might be able to identify the minimum level of *disposable income increase* that *households consider positive*.

We have already touched upon the difference between "statistical significance" and "economic significance" (Engsted, 2009), mentioning that "economic significance" refers to the practical, real-world implications of statistical results distinguished from the purely statistical findings. I would draw an analogy here with "clinical significance": the real question behind "economic significance" is the practical or real-world importance of statistical findings in analogy to "clinical significance." In other words, "economic significance" looks at the size of the effect in terms of its economic impact, policy implications, or practical outcomes.

The concept of "economic significance" prevents researchers and stakeholders from drawing unwarranted conclusions based purely on statistical results ("statistical significance") without considering the real-world importance of their intervention or program. Two illustrative questions could be in this context as follows: Is the effect large enough to influence and guarantee a decision or policy? Does the effect have any real-world implications that people would perceive?

For example, when it comes to the unemployment rate in India, the Centre for Monitoring Indian Economy (2024), abbreviated as CMIE, also underscores that it is essential to determine what levels of change in the unemployment rate would trigger meaningful policy interventions. However, the CMIE (2024) report might be critiqued for not specifying what constitutes a meaningful/substantial change in unemployment that would warrant government action. For example, *policymakers need to know whether, for instance, a 0.5% change in the unemployment rate is substantial enough to implement new employment schemes*.

Moreover, the CMIE data might show fluctuations in unemployment rates over time. From an MID perspective, it is critical to distinguish between short-term variability and long-term trends that have a sustained impact on economic and social policies. *Short-term fluctuations might be less practically significant than long-term trends indicating a systemic issue in the economy*. But let this issue rest for the time being.

Card and Krueger (2000) already examined the impact of increases in minimum wage, using economic thresholds for meaningful/substantial change that resonate well with the MID/MCID concept in clinical research. The two authors, for example, found no indication that, despite the *rise in the minimum wage*, there was a *reduction in employment*.

A very important point of development came from Angrist and Pischke (2009). They proposed the widespread differences-in-differences (DiD) model and tried to establish thresholds of change (the so-called "impact size").

DiD is a statistical procedure, very similar to RCTs in the clinical field in my opinion, that compares the outcomes of a treatment group to a pre- and posttreatment control group (or policy change, if it is about the economy, for example). DiD is a common technique in the evaluation of studies in economy, intervention, or treatment regarding their impact and it is employed in diverse fields, including health research, economics, or public policy (e.g., Angrist & Pischke, 2009).

DiD is a quasi-experimental econometric technique used to estimate causal relationships when controlled experiments are not feasible as in psychology or the clinical field, for example. The method is primarily applied in policy evaluation studies to determine the "treatment effect" of a policy or intervention by comparing the changes in an outcome over time between a group that is exposed to the treatment and another group that is not exposed to it (Angrist & Pischke, 2009).

The standard DiD model typically involves data from two periods (pretreatment and posttreatment) and two groups (treated and control). In the pretreatment period, both groups are assumed to be following similar outcome trajectories, that is, they are exposed to the "treatment." After the intervention, if the treated group's outcomes diverge from the control group in a way that exceeds any natural divergence (such as negligible statistical noise) expected without treatment, this difference is attributed to the effect of the treatment.

In DiD, the "two differences" refer to comparing the change in an outcome between two groups, both before and after the intervention is performed. Essentially, the difference in the outcome change between the treated group and the control group is calculated; this shows the effect of the treatment by controlling any preexisting differences between the groups.

Key points about the two differences: The first difference is the difference in the outcome (similar to PRO in the clinical field) between the first group and the second group before the intervention, while the second one is the difference in the outcome between the first group and the second group after the intervention (post-measurement).

By subtracting the "first difference" from the "second difference," we can isolate the impact of the intervention, as it removes any preexisting differences between the treatment and control groups that might have influenced the outcome. An important assumption is that for the DiD method to be valid, it is crucial to assume that both the treatment and control groups would have experienced similar trends in the outcome variable over time if the intervention had not been implemented (*parallel trends assumption*).

Also, I would add that the baseline characteristics and the temporal trajectories should be comparable. Just think of clinical/medical trials in which the two groups should be comparable on the relevant variables.

Angrist and Pischke (2009) typically specify the DiD model as a linear regression with interactions between a time dummy (pre/post) and a treatment dummy (treated/control). This allows them to isolate the effect of treatment on the treated group after accounting for time trends common to both groups.

Angrist and Pischke (2009) also emphasize checking for violations of the *parallel trends assumption* earlier mentioned and considering the inclusion of covariates to control for observable characteristics (earlier mentioned as baseline characteristics).

However, Angrist and Pischke (2009) caution that the method relies heavily on the assumption of parallel trends, which must be justified through careful examination of the data. The primary strength of DiD, as noted by them, is that it helps control unobserved confounders that do not vary over time, making it a widely used tool for causal inference in policy evaluation. Unfortunately, in the clinical field, confounders can vary over time, for example, the functional status of the patient.

Still, I believe that the DiD method does not address the question of MID properly and the individualized characteristics either. I should think that DiD could benefit a lot from the MID framework, especially when it comes to economic interventions. In those cases, anchor-based methods could be used to determine the MID.

I can see another problem with the DiD model by Angrist and Pischke (2009), namely, that they work with linear regression. In many cases, trends over time are nonlinear, which calls for a nonlinear regression model. They even go as far as claiming to be able to prove causality. However, given the previously mentioned criticism (nonlinearity) and interindividual differences, this statement seems too strong a scientific claim.

My criticism about linear regression is in accordance with Hanff et al. (2024) who argue against the use of t-tests or linear regression in the case of apathy in Parkinson disease. I find it conceivable that the criticism by Hanff et al. (2024) applies to other diseases and not just to Parkinson disease.

I would like to praise Angrist and Pischke (2009) for including an MID-like metric. Discontinuity design (Regression discontinuity design or RDD), one of the quasi-experimental methods, helps us estimate causal impacts in cases where random assignment is not possible. It takes advantage of a cutoff or threshold above or below which treatment is assumed to differ, similar to MID. The assumption for RDD is that people right above and right below the cutoff are very similar, apart from the treatment (intervention) measures. It is a shift in the probability of receiving treatment at some value (cutoff) on the continuous scale which can be treated as discontinuity.

Admittedly, RDD is capable of capturing policy strategy evaluation and assessing policies that utilize cutoffs. However, I could identify some limitations. For example, in the education domain, estimating the impact of scholarships or remedial classes offered to students who qualify based on standardized test results might be unfair. In the field of health policy, evaluating the effects of treatment provided at certain clinical borderline thresholds can be debated: for example, payment by cholesterol level based on a cutoff.

To combine health with economics, I would like to cite Brouwer and Bateman (2005) whose seminal study deals with determining the thresholds at which health-risk reductions are valued by different populations. These thresholds are essential for policy decisions, similar to how MCIDs or MIDs are used.

Their concept of WTP is analogous to MID in that both aim to quantify a threshold of change that is considered practical or significant to the client. In economic studies like Brouwer and Bateman's (2005), WTP estimates help determine the *economic value individuals place on risk reductions*. WTF refers to the monetary amount that people are willing to pay to decrease the likelihood of experiencing negative health outcomes such as a disease or a disaster (flood).

This value helps policymakers and economists understand how much individuals value safety and health improvements, which can guide resource allocation and investment in public health measures, and also stakeholders when it comes to tagging the price of a product.

A simple numerical example could be that without medicine, the risk of contracting disease is 10%, while the cost if the disease is contracted equals $1000 (medical expenses, lost income, transportation, hospitalizations, etc.). In contrast, with the medicine, the risk of contracting the disease is 1%, while the cost if the disease is contracted is $100. The reduction in risk (from 10% to 1%) is valued by individuals at $50. This WTP amount reflects the value they place on reducing their risk of facing the high costs and negative consequences associated with the disease.

This WTP information is valuable for Public Health Policy in that governments can use this data to decide whether to subsidize or fund vaccination programs, for example. Also, Health economists can compare the costs of implementing the vaccination program with the total WTP amounts to determine if the program is economically justified.

The article's approach to comparing WTP estimates across countries can be linked to the variation in MCID thresholds across different populations or cultural contexts. Just as WTP values might differ due to socioeconomic, cultural factors, and beliefs, the perception of what constitutes an MID in health outcomes may also vary across different demographic groups, as discussed earlier.

The integration of WTP estimates into health economics models aligns with the use of MCID/MID in clinical trials to evaluate the cost-effectiveness of interventions. Both approaches aim to make informed decisions that maximize health benefits within the constraints of available resources.

In summary, Brouwer and Bateman's study (Brouwer & Bateman, 2005) on WTP for health-risk reductions provides valuable insights into how economic valuations can inform health policy, similar to how MCID/MID thresholds inform clinical decision-making. The willingness to pay for health-risk reductions differs across different countries, which again supports my claim that any MID or MCID should be individualized. In the present case, not at the individual level, of course, but at the national level. Crucially, the maximum set of variables should be considered to get a precise estimate of MID, be it in the context of WTP or a medical condition.

For instance, when assessing the impact of a government stimulus package, policymakers can establish MIDs for key economic indicators such as GDP growth, unemployment rates, and consumer spending. By measuring changes against these MIDs, policymakers can determine whether the stimulus has achieved meaningful economic improvements.

Within environmental economics, MID also assists in assessing the effectiveness of policies aimed at mitigating environmental damages. For instance, take a hypothetical government program designed to lower carbon emissions using a cap-and-trade system. With set MIDs for emission reductions, the policymakers can measure whether the program is meeting its environmental targets. MIDs can be used to redirect policies and hence can significantly reduce greenhouse gas emissions.

Moreover, MIDs can also guide investment decisions and risk assessments in finance. MIDs often serve as benchmarks used to measure the success of investment portfolios. For example, when determining the effect of some market movements on a portfolio's value, analysts can set MIDs that correspond with certain levels of risk and reward which are acceptable. This enables the management of assets and risks.

MID's practical use cases in economics and finance are exemplified further in real-life scenarios. During the financial crisis of early 2008, governments around the globe tried to execute several stem measures to stabilize the economy. With the set MIDs for unemployment and GDP growth, the policymakers evaluated how effective these interventions were. The US government's American Recovery and Reinvestment Act sought to create or save several millions of jobs. MIDs were created to assess the outcome of these activities.

The concept of MID is also relevant in estimating the impact of green investments in environmental finance. For instance, financing of renewable energy projects often comes with the evaluation of MID for the emission reductions that are incurred as a result of these investments. Reis et al. (2023) pointed out the relevance of setting "MIDs" (although not explicitly mentioning the notion) for low-carbon investments within the context of climate objectives emphasizing how financial choices can support environmental goals.

To sum up, the use of MID in economics, economic policy, environmental economics, and finance serves important purposes for stakeholders, investors, and researchers. MID sets reasonable limits and terms for evaluating alterations in economic parameters and the socio-environmental impacts, thus improving the policy and investment allowance effectiveness. With the perpetual changes in the world economy, the incorporation of MID in economic and financial policymaking will be critical for achieving sustainable development alongside other environmental issues.

Health Economics

Within health economics, the concept of MID relates to the interpretation of cost-effectiveness analyses. For example, consider the interface between economics and health care. The idea is to balance and navigate optimal resource allocation toward *maximization of benefits at a minimum possible cost*. This particular aspect is very relevant within the domain of health economics when trying to achieve more health outcomes in relation to expenditure is the main objective. Primary care physicians can use MID to frame the value of interventions to economic evaluations and hence make better decisions toward resource distribution.

MID has various implications in health care, specifically in surgical procedures. Rodrigues et al. (2015) studied the MID in elective hand surgery and postulated how defining meaningful change helps in increasing patient satisfaction, implying that there is considerably greater appreciation of the notion. This is especially important within the realm of economic evaluations because the estimated worth of surgical procedures affects funding and the availability of resources as well as their uses in the health systems.

The incorporation of MID into health economics is also justified by the advancement of literature on patient centeredness in health care. This is in tandem with the shift to value-based care, where health systems are progressively trying to achieve what is important to patients, not just what can be measured clinically.

In sum, the notion of MID emerges as a critically important measure in achieving the synthesis of clinical results and economic analysis in health care. By prioritizing patient perceptions and values, MID sheds light on what is an important change in health status. MID is crucial in making health-care decisions, managing resources appropriately, and enhancing PROs. As health economics expands, incorporating MID in economic evaluations will be imperative to making sure that health-care systems are attentive to the patient's needs and preferences.

The scope of MCID and MID applies to health economics, particularly in cost-effectiveness analysis (CEA). The metric assists in determining whether or not the treatment outcomes justify the costs by measuring if the benefits of a treatment are visible and meaningful from the patient's perspective (Terwee et al.). This is especially critical in developing countries.

A clear and simple example concerning the use of some antimalarial medications illustrates the impact of resource limitations on health care. Consider a region with high incidence of malaria and with poorly developed health-care systems. There is a high prevalence of the use of antimalarial medications such as Artemisinin-based combination therapies (ACTs). In this situation, MCID would be very important in determining the effectiveness of different ACTs on reducing malaria parasitemia. For example, let us say a clinical trial compares two ACTs, Artemether-Lumefantrine and Dihydroartemisinin-Piperaquine, on their effectiveness in decreasing parasitemia. The MCID here would assist in estimating the minimum decrease in parasite count that would be considered beneficial to patients. If a drug achieves a decrease in parasitemia that meets or exceeds the predetermined MCID, then it would be deemed more effective for that particular outcome.

In the case of chloroquine, which is an older antimalarial, it is in areas where plasmodium falciparum has developed considerable resistance that the drug would less likely eliminate parasitemia or avert complications including morbidity and mortality. This scenario could be viewed as not attaining minimal clinically important difference (MCID) of effectiveness.

Let me end with a citation by Michael Schlichting et al. (2022) who was already skeptical in 2022: "the $\geq$ 15% response threshold is incongruent with previously defined and scientifically established thresholds and is not well-suited for universal implementation. Further scientific evidence and discussion among all stakeholders are needed, especially should this universal rule be considered in the context of future joint clinical assessments of health technologies in the European Union scheduled from 2025 onward."

References

Aleluia Reis, L., Vrontisi, Z., Verdolini, E., Fragkiadakis, K., & Tavoni, M. (2023). A research and development investment strategy to achieve the Paris climate agreement. *Nature Communications, 14*(1), 3581. https://doi.org/10.1038/s41467-023-38620-4

Angrist, J. D., & Pischke, J. S. (2009). *Mostly harmless econometrics: An empiricist's companion*. Princeton University Press. https://doi.org/10.1515/9781400829828

Brouwer, R., & Bateman, I. J. (2005). Benefits transfer of willingness to pay estimates and functions for health-risk reductions: A cross-country study. *Journal of Health Economics, 24*(3), 591–611. https://doi.org/10.1016/j.jhealeco.2004.09.004

Card, D., & Krueger, A. B. (2000). Minimum wages and employment: A case study of the fast-food industry in New Jersey and Pennsylvania: Reply. *The American Economic Review, 90*(5), 1397–1420. http://www.jstor.org/stable/2677856

Centre for Monitoring Indian Economy. (2024). *Unemployment rate in India*. CMIE. Retrieved [January 04, 2024], from https://www.cmie.com/kommon/bin/sr.php?kall=warticle&dt=20240104181058&msec=536

Engsted, T. (2009). Statistical vs. economic significance in economics and econometrics: further comments on McCloskey and Ziliak. *Journal of Economic Methodology, 16*(4), 393–408. https://doi.org/10.1080/13501780903337339

Hanff, A.-M., Krüger, R., McCrum, C., & Ley, C. (2024). Mixed effects models but not t-tests or linear regression detect progression of apathy in Parkinson's disease over time: A comparative analysis. *BMC Medical Research Methodology*. https://doi.org/10.17605/OSF.IO/NF4YB

Rodrigues, J. N., Mabvuure, N. T., Nikkhah, D., Shariff, Z., & Davis, T. R. (2015). Minimal important changes and differences in elective hand surgery. *The Journal of Hand Surgery, European Volume, 40*(9), 900–912. https://doi.org/10.1177/1753193414553908

Schlichting, M., Hennig, M., Rudell, K., McLeod, L., Bennett, B., Shaw, J., Doward, L., Molsen-David, E., & Chassany, O. (2022). Is IQWiG's 15% threshold universally applicable in assessing the clinical relevance of patient-reported outcomes changes? An ISPOR Special Interest Group Report. *Value in Health: The Journal of the International Society for Pharmacoeconomics and Outcomes Research, 25*(9), 1463–1468. https://doi.org/10.1016/j.jval.2022.07.010

The manufacturer's authorised representative in the EU is Springer Nature Customer Service Centre GmbH, Europaplatz 3, 69115 Heidelberg, Germany. If you have any concerns regarding our products, please contact ProductSafety@springernature.com

Printed and bound by CPI Group (UK) Ltd, Croydon, CR0 4YY

05/10/2026

02228129-0002